Vision Sensors and Edge Detection

Editor

Bilroy Muller

scitus academics

Vision Sensors and Edge Detection
Edited by **Bilroy Muller**

ISBN: 978-1-68117-588-1
Library of Congress Control Number: 2016946443

SCITUS Academics LLC,
616, Corporate Way, Suite 2, 4766,
Valley Cottage, NY 10989

www.scitusacademics.com

Notice

Reasonable efforts have been made to publish reliable data and views articulated in the chapters are those of the individual contributors, and not necessarily those of the editors or publishers. Editors or publishers are not responsible for the accuracy of the information in the published chapters or consequences of their use. The publisher believes no responsibility for any damage or grievance to the persons or property arising out of the use of any materials, instructions, methods or thoughts in the book. The editors and the publisher have attempted to trace the copyright holders of all material reproduced in this publication and apologize to copyright holders if permission has not been obtained. If any copyright holder has not been acknowledged, please write to us so we may rectify.

Printed in United States of America on Acid Free Paper

Preface

Vision Sensors/Machine Vision Systems analyze images to perform appearance inspections, character inspections, positioning, and defect inspections. Vision sensors are products consisting of a video camera, display and interface, and computer processor to automate industrial processes and decisions. These are commonly used for measurement, pass/fail decisions, and other observable characteristics relating to product quality. When the camera has an integral processor, it is called a smart camera. Edge detection is an image processing technique for finding the boundaries of objects within images. It works by detecting discontinuities in brightness. Edge detection is used for image segmentation and data extraction in areas such as image processing, computer vision, and machine vision.

Vision Sensors and Edge Detection provides a comprehensive overview of recent developments within the area of vision sensors and edge detection. It emphasizes on vision sensors with applications to panoramic vision sensors, wireless vision sensors, and automated vision sensor inspection, and image processing techniques.

Table of Contents

Chapter 1

Design of Omni-Directional Tilt Sensor Based on Machine Vision

Yiping Tang[1], **Caiguo Chen**[2]

[1] College of Information Engineering, Zhejiang University of Technology, Hangzhou, China

[2] College of Computer Science, Zhejiang University of Technology, Hangzhou, China

ABSTRACT

This paper presents a new Omni-Directional Tilt Sensor (ODTS), which consists of the LED light, transparent cone-shaped closed container, mercury, camera, embedded systems and so on. The volume of the mercury in the container is equal to half of the container's. When the detected surface is horizontal, the shape of mercury in the image captured by the camera is a black disc since the mercury is lightproof. When the detected surface tilts, the mercury flows and the mercury surface always maintains horizontally due to the gravity force of the earth. At this time, some area of the transparent cone-shaped closed container is not shaded by mercury and the border of the mercury's shape in the captured image is a half circle and a half ellipse. Thus there is a translucent crescent-shaped area in the image. With analyzing this area by the specific algorithm based on machine vision, the tilt angle and directional angle can be obtained. The experimental results show that the ODTS proposed has some advantages, such as simple maintenance, high precision, wide range, low cost, real-time, reliability and high visualization.

KEYWORDS

ODTS, Machine Vision, Digital Image Processing, Embedded System, Tilt Angle and Tilt Directional Angle

1. INTRODUCTION

Tilt sensor is mainly applied to measuring the horizontality of a system and widely used in construction of bridges, laying of railway, civil engineering, oil drilling, aviation and maritime, industrial automation, intelligent platform, machining and other important fields. There are several kinds of commercial sensor for tilt angle measurement, which are available in the market. It mainly can be divided into three kinds solid pendulum [1,2], liquid pendulum [3] and gas pendulum [4] based on different working principles. They all have their advantages and disadvantages in terms of their working principles. In the gravitational field, the sensitive medium of solid pendulum is the weight of pendulum, while liquid pendulum's is electrolyte and gas pendulum's is gas. With a lot number of applications, the research of tile sensors of solid pendulum has been quite mature, but it is easy to be interfered by the external factors, such as the vibration caused by mechanical shock. The tilt sensor of liquid pendulum, however, has many advantages, such as high sensitivity, corrosion and moisture resistance, etc. But its fatal weakness is that its working performance is easily affected by the change of temperature and this fatal weakness limits its development and applications. The tilt sensor of gas pendulum has a simple structure, strong resistance to vibration and shock, but it can be greatly affected by environmental temperature and is low in measuring precision. In addition, some experts proposed a method to detect tilt angle according to the images [5]. However, it can only get the one-dimensional tilt angle because only the position of bubbles is detected.

The literature [6] has proposed a vertical-pendulumbased fiber Bragg Grating tilt sensor, which can detect the magnitude as well as the direction of the inclination from the horizontal direction. Fiber Bragg Grating (FBG) sensors is based on the modulation of the reflection wavelength of FBGs in response to temperature and strain. Compared to other types of fiber-optic sensors, FBG sensors have advantages of being inherently self-referencing and the capability in multiplexing a large number of FBG sensing elements. The sensor encodes the tilt angle with the wavelength separation between two FBGs to overcome the problem of temperature cross-sensitivity. Also it has high accuracy and accuracy of 0.1°and resolution of better than 0.007°can be achieved. But it is hard in maintenance and is not an optical system. Thus it can't give us direct visual results.

The literature [7] has proposed an optical sensor including an opto-electronic sensor with laser, optical components and image sensor. Its working principle is based on optical interferometry. And it has high measurement accuracy and is insensitive to magnetic environment. Also it can measure two axes at the same time. But because the liquid medium is oil and it can wet the glass, the quality of captured image may be affected thus lead to the measurement error.

Here we propose the Omni-Directional Tilt Sensor (ODTS) based on machine vision. In terms of tilt perception, it belongs to the tilt sensor of liquid pendulum. Different from existing tilt sensor of liquid pendulum, the liquid medium of ODTS is mercury whose advantages include avoiding wetting the glass and plastic, light-proof, good conductivity, etc. It minimizes the temperature interferences to the ODTS and can be used in the environment of large temperature range. Also the energy saving design is added to the sensor using the conductivity of mercury. The tilt angle and tilt directional angle is obtained by analyzing the image formed by the mercury in the container in the way of digital image processing. Thus it avoids some indirect measurement segments that may result in unnecessary measurement error. A prototype sensor is designed and experimental results show the tilt angle relative to the sea level can be detected at the accuracy of 1°and at the resolution of 0.01°

2. DESIGN OF THE ODTS

The ODTS based on machine vision consists of shell of the ODTS, LED light, transparent cone-shaped closed container, mercury , camera , embedded systems, power supply and compass, showed in **Figure 1** ("1" is the shell of the ODTS, "2" is the LED light, "3" is the transparent cone-shaped closed container, "4" is mercury, "5" is the camera, "6" is the embedded system, "7" is the power supply, "8" is the compass). The power supply provides the LED light and the embedded systems electricity. Then the embedded systems offer electricity to the camera through USB interface. And the transparent coneshaped closed container is composed of two cones with the same size and shape in the way back to back. It is fixed in the middle of the shell of ODTS and is always parallel with the detected surface, above of the container is the LED light and below is the camera. The LED light, facing right down to the transparent cone-shaped closed container, projects the white light. The camera captures pictures of the transparent cone-shaped container. At the same time, the embedded system reads the data of pictures from the camera through the USB.

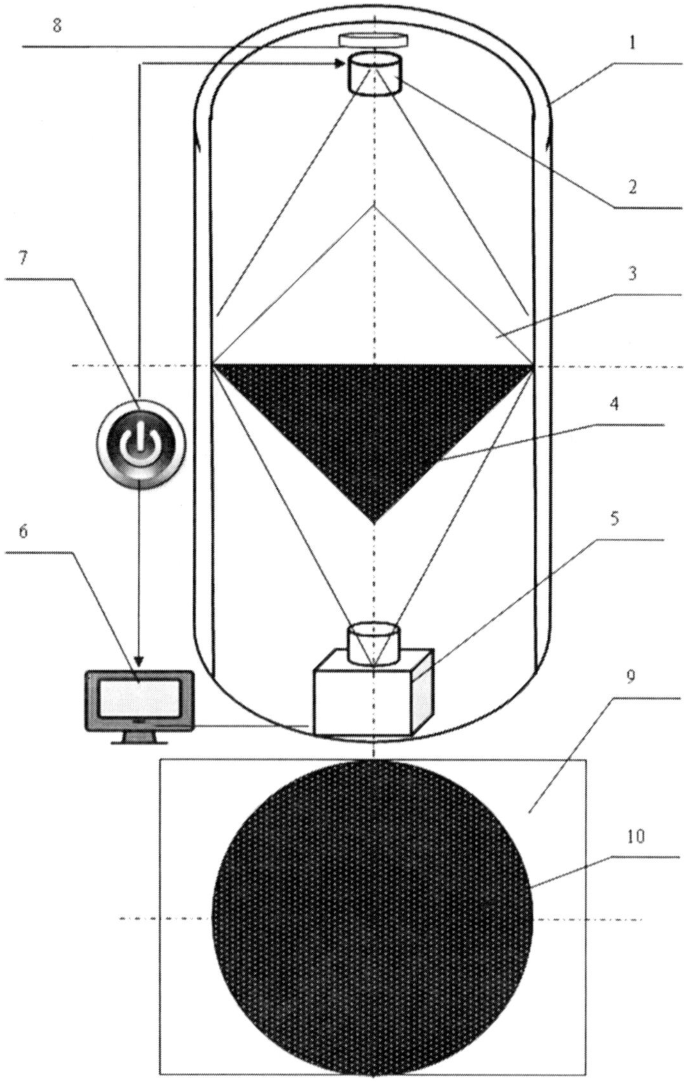

Figure 1. Structure of ODTS.

2.1. Design of the Transparent Cone-Shaped Container

When assembling the transparent con-shaped container, the mercury is poured into it and the volume of mercury is half of the container's. According to physical principle, when the ODTS is in the level state, because the light is shaded by the mercury in the transparent cone-shaped container, the camera can't catch the transmitted light penetrating the container from the LED light, showed in **Figure 1** ("9" is the captured image by camera, "10" is the image of the transparent cone-shaped closed container). When the surface to be detected tilts, the mercury flows and the mercury surface always maintains horizontally due to the gravity force of the earth. Because the transparent cone-shaped container is composed of two cones with the same size and shape, the volume of mercury flowing out of one cone is equal to the volume of mercury flowing into the other one and the level surface always passes the central point of the transparent coneshaped container. In other words, the level surface of mercury in the state of tilt rotates around Y axis. Thus some area of the transparent cone-shaped container between the LED light and the camera is in the state of non-shelter, showed in the **Figure 2**. From the capturing angle of camera, the shape of mercury is a disc in the projection image if the tilt condition doesn't happen, however, it would be a half-circle and a half-ellipse if the tilt condition happens. The major axis of ellipse is equal to the radius of disc and the tilt angle has functional relationship with length of the minor axis of ellipse. The smaller the minor axis of ellipse is, the bigger the tilt angle is. And the tilt directional angle is in the negative direction of ellipse's minor axis.

The resolution of the tilt angle is related to the radius R and the cone angle α of the transparent cone-shaped container. The equation of their relation is below.

$$\theta = ctg^{-1}\left[\left(R/\delta - 1\right)\times ctg\left(a\right)\right]. \tag{1}$$

In the equation, R is the radius of the transparent coneshaped container, α is the cone angle of it, δ is the middle width of the crescent-shaped translucent part, θ is the tilt angle.

According to the Equation (1), the longer the radius R of the transparent cone-shaped container is, the higher the resolution of tilt angle θ is, and the resolution of tilt angle θ has functional relationship with the cone angle α. Generally speaking, the pixels of the transparent coneshaped container's radius R in the captured image is up to the vision range of camera and in this paper, the radius R of the transparent cone-shaped container is 38 mm and possesses 188 pixels in the captured image. Also the cone angle α is determined by the actual level measurement range. The top of **Figure 3** is several cones with different angles, the bottom of it is the curve diagram of the relationship between the middle width δ of the crescent-shaped translucent part and the tilt angle θ under the condition that the radius R of the transparent cone-shaped container is 38mm and the cone angle α is 15˚, 30˚, 45˚, 60˚ and 75˚. **Figure 3** shows that the small tilt angle θ has high resolution when the cone angle α is 15˚. Therefor the choosing or design of the cone angle α of transparent cone-shaped container is determined by the actual measurement precision.

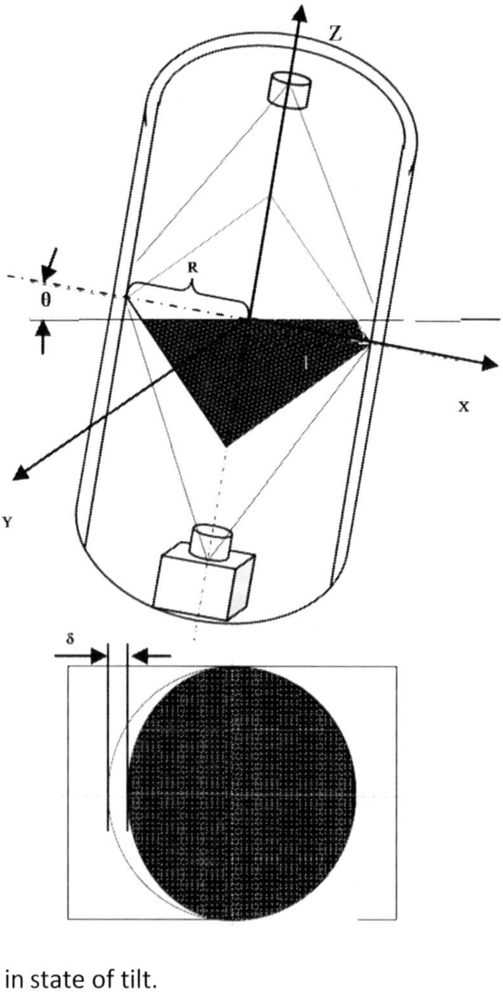

Figure 2. Image in state of tilt.

2.2. Liquid Medium for ODTS

Mercury is suitable to be the liquid medium in the ODTS with the advantages of not wetting the glass and plastic, light-proof, good conductivity, etc. Because mercury keeps liquid state in a large temperature range (–38~ + 356.66°C), the ODTS with mercury as the liquid medium has great adaptability of the environmental temperature. Besides, it can avoid the interference to the tilt measurement caused by the vibration in the vertical direction with the strong surface tension and large proportion of mercury and is able to makes the ODTS get the tilt angle and directional angle accurately using the light-proof of mercury. Besides, the electrode is connected when the tilt sensor leans to one side with the good conductivity of mercury, which can be applied to power saving. That means it activates the embedded system and connects the power supply for camera and LED light when the tilt condition happens. And the power supply is off and the embedded system is in dormant mode when the tilt condition doesn't happen. It is known the thermal expansion coefficient of mercury is 1.8×10^{-4} °C, in other words, if the temperature changes one degree, the volume of mercury changes 1/5556. According to the volume calculating equation of the cone, we calculate the derivative of temperature varying for the volume and get the Equation (2).

$$dV/dT = \pi R^2 \tan(\alpha) dR/dT. \qquad (2)$$

In the equation, dV/dT is the thermal expansion coefficient of mercury. Combining with the Equation (1) and calculating the derivative, the change of tilt angle θ caused by temperature's change would be attained. According to the tilt detecting method in this paper, the measurement error caused by the temperature change can be controlled in within the scope of 1/1000.

2.3. Software Design for ODTS

The calculating part of the ODTS includes the embedded hardware and embedded software including system software and user software. Because of the paper's length limit, only the user software will be discussed in the paper. The user software consists of image obtaining unit, tilt angle and tilt directional angle detecting unit, system data storing unit, detecting data storing unit, displaying unit of the detecting result. And the overall block diagram of system software is showed in **Figure 4**. The image obtaining unit, including system initialization module and image reading module, is mainly used to read the video data from camera. The initialization module of the system is used to read some system constant which are stored in system data memory unit such as the radius R and the cone angle α of the transparent cone-shaped container, initial directional angle β0, resolution of the camera, the calculating table of tilt angle θ and the width δ of the translucent part, etc. And the image reading module is mainly used to read the video data from camera and save them in the dynamic memory unit. The tilt angle and tilt directional angel detecting unit is used to detect and calculate the tilt angle θ and tilt directional angle β of the detected object, including tilt directional β detecting module, tilt angle θ detecting module, the calculating module of the change rate of tilt angle θ and tilt directional angle β.

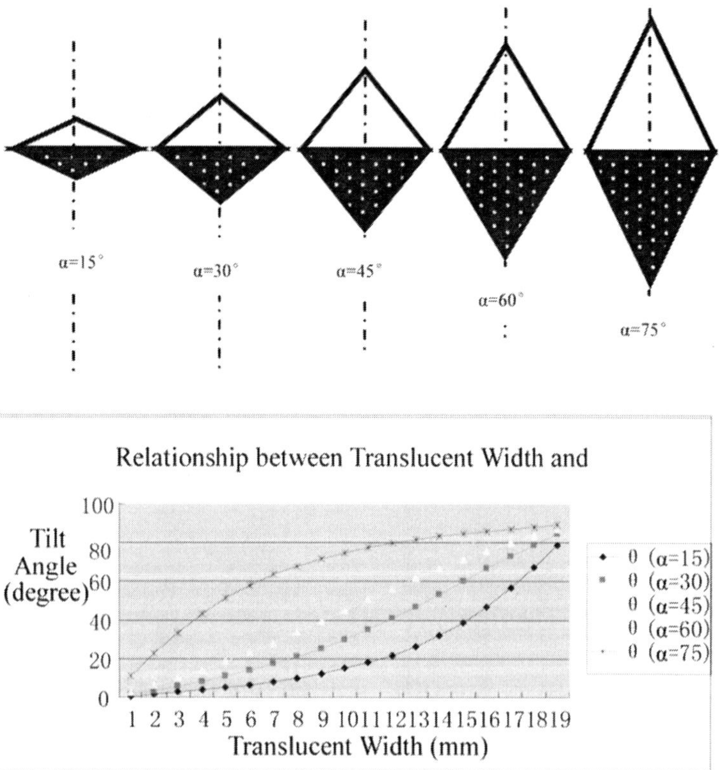

Figure 3. The curve between the cone angle α and the resolution of tilt measurement.

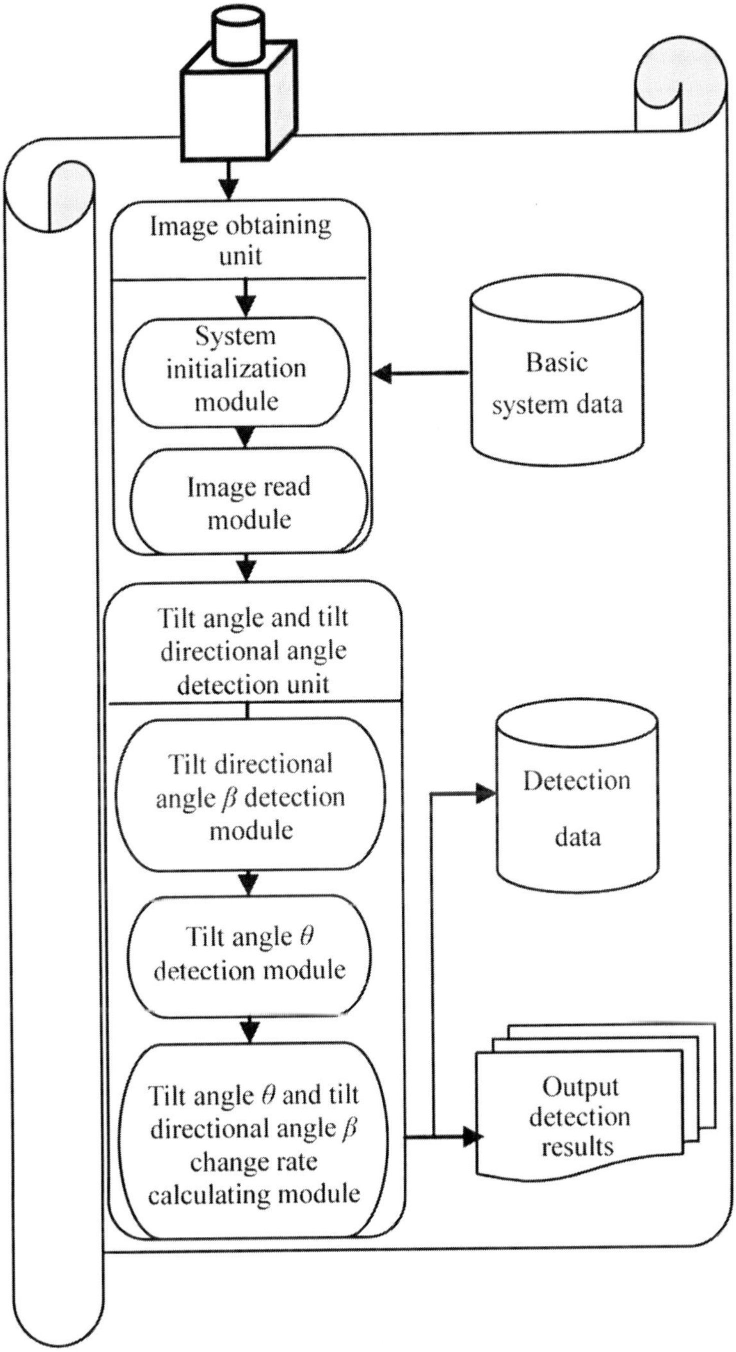

Figure 4. Overall block diagram of system software.

2.4. Detecting Algorithm for the Directional Angle β of ODTS

The definition of tilt directional angel expressed in angle β in this paper is a clockwise angle started from north direction. The tilt directional angle detected in the image is clockwise from the X positive axis, and it is expressed as angle βx, showed in **Figure 5**. Thus there is a following relationship between tilt directional angle β and detected tilt directional angle βx, showed in Equation (3).

$$\beta = \beta x + \beta 0. \tag{3}$$

In this equation, β is the tilt directional angle, βx is detected tilt directional angle, β0 is initial directional angle.

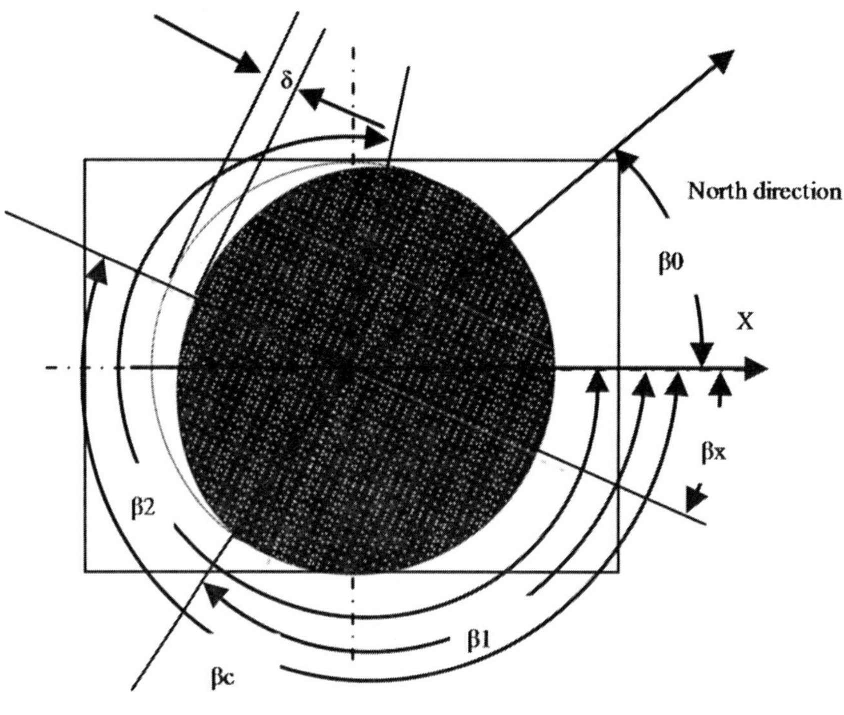

Figure 5. Detecting algorithm of tilt angle and tilt directional angle.

The initial directional angle $\beta 0$ is determined by the angle between the line of the cylindrical outer wall and the X axis's direction of obtained video image during the ODTS's calibration test, and it is written into the system data memory unit. The detected tilt directional angle βx is calculated based on the geometry of translucent part in the captured image, shown in **Figure 5**. It is known that the tilt directional angle is in the negative direction of ellipse's minor axis mentioned above. And the location of the minor axis would be in the middle of the crescentshaped's part. The detecting algorithm is showed as follows:

Step 1: generate a line in X axis's positive direction and search around the center point of the image in clockwise direction, if there is no bright pixel in X axis's positive direction, then search in the way mentioned above, if there is, skip to the step 3, if the circular outer pixel which the rotating line meets is bright pixel, the angle between the rotating line and the X axis's positive direction is $\beta 1$.

Step 2: continue searching around the center point in clockwise direction, if the circular outer pixel which the rotating line meets is not bright pixel, the angle between the rotating line of last bright pixel and the X axis's positive direction is $\beta 2$, then skip to step 5.

Step 3: continue searching around the center point in clockwise direction, if the circular outer pixel which the rotating line meets is not bright pixel, the angle between the rotating line of last bright pixel and the X axis's positive direction is $\beta 2$.

Step 4: continue searching around the center point in clockwise direction, if the circular outer pixel which the rotating line meets is not bright pixel, the angle between the rotating line of last bright pixel and the X axis's positive direction is $\beta 1$.

Step 5: use Equation (4) to calculate the angle βc of the ellipse's minor axis;

$$\beta c = (\beta 1 + \beta 2)/2 \tag{4}$$

And the detected tilt directional angle βx must be in the negative direction of ellipse's minor axis, therefor the detected tilt directional angle βx can be expressed in Equation (5),

$$\beta x = \beta c - \pi \tag{5}$$

2.5. Detecting Algorithm for the Tilt Angle θ of ODTS

The tilt angle is calculated with the Equation (1), and the parameters such as the radius R of the transparent coneshaped container, the cone angle α of the transparent cone-shaped container and so on, are attained from system data memory units as system constant. The width δ of the translucent crescent-shaped's middle part is calculated with the analysis algorithm of the image, specific algorithm is followed:

The pixel number of the translucent crescent-shaped's middle part is calculated according to the angle βc of the ellipse's minor axis resulting from the Equation (4) and the translucent crescent-shaped area which is formed by the transmitted light. If the resolution of the camera is 640 × 480, and the radius R of the transparent cone-shaped container is 200 mm, every pixel represents 0.83 mm. If the calculated pixel number of the translucent crescentshaped's middle part is five pixels, the width δ of the translucent crescent-shaped's middle part is 4.15 mm.

3. REALIZATION OF THE ODTS

According to the above-mentioned design thought, the prototype of the ODTS in this paper is designed, shown in **Figure 6**. To avoid the detecting interference from the environmental light, the transparent cone-shaped container, camera, and LED are fixed in a closed cylinder so that the ODTS can be used even in the harsh environment.

The transparent cone-shaped container filled with mercury whose volume equals to half of the container's is composed of two opposite faced cones with the cone angle of 15° showed in **Figure 6**(a).

According to the software block diagram showed in **Figure 4**, we have developed the application-oriented software of the ODTS, showed in **Figure 7**. With the aim to confirm the feasibility of the prototype and the application-oriented software, we had experimental study for the prototype on the platform of Windows XP/2000, JAVA (JDK1.5 version) system development language, Java Media Framework 2.1.1e video multimedia-based software.

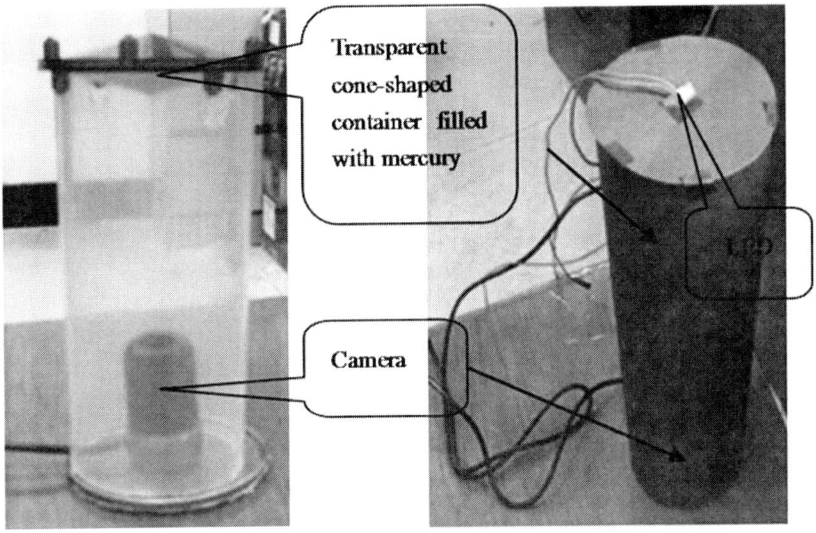

(a) (b)

Figure 6. Prototype of ODTS.

In order to make the software versatile and more userfriendly in various environments for users, the system is designed to have four custom-designed windows: 1) custom-designed window of video data source; 2) customdesigned window of system parameter; 3) custom-designed window of control parameter; 4) custom-designed window of detection area.

The custom-designed window of video data source is used to determine the obtaining devices of the tilt image video data, considering there are many interfaces of image sensor, such as USB video data interface, IP video net interface, simulation video interface, and AVI video data. The custom-designed window of system parameter is used to determine the radius R and cone angle α of the transparent cone-shaped container and to make the software more flexible. The custom-designed window of control parameter allows the users to carefully observe the detecting process of the tilt and provide real-time detection data for artificial leveling. The custom-designed window of detection area is used to customize the video source detection area. Because the body part of the disc of the image captured by camera is not in the center of image for sure, customizing the detection circle manually is needed to eliminate all kinds of errors in the process of assembling the ODTS.

4. EXPERIMENTAL RESULTS AND DISCUSSION

The ODTS is placed on the detected surface and is parallel with the detected surface. When the ODTS is horizontal, the shape of mercury in the image captured by the camera is a black disc and the RGB value of pixel is (0, 0, 0) which means black. At the same time, the RGB value of remaining pixels in the image is (255, 255, 255) which means white showed in**Figure 8**. The captured image is 640 × 480 pixels. With the Hough circle detecting algorithm based on diameter [8], the detected center of the black disc in the image was (331,267), and the radius of which possessed 188 pixels. The experimental research for the prototype showed in the**Figure 6** was done under the condition that the cone angle α of transparent coneshaped container was 15°, the actual radius R of the transparent cone-shaped container was 38mm, the light source was a white LED, and the resolution of camera was 640 × 480 pixels. So every pixel in the image represented 38/188 mm.

When the ODTS tilted 7° artificially, the formed shape of mercury in the image and the detected results are showed in **Figure 9**. With the detecting algorithm of tilt directional angle, the angle βc of the ellipse's minor axis was obtained, then the pixel number of the translucent crescent-shaped's middle part was calculated and the number was 59. Because the width of each pixel in the image was known, so the middle width δ of the crescent-shaped translucent part was 11.9 mm. With the Equation (1) and the known parameters of R, α and δ, the detected tilt angle θ was 6.98°. Also we have done some experiments under the condition the ODTS tilted 9°, 12° and 15° artificially, and the detected results are showed in **Table 1**.

By analyzing the table, the detected error of each tilt angle is within ± 1°, and the relative error is within ±5%. The detected error is relatively small when detecting a small tilt angle. The reason for that is the resolution of the ODTS with a small cone angle reacts sensitively when detecting a small tilt angle. It is consistent with the earlier analysis above.

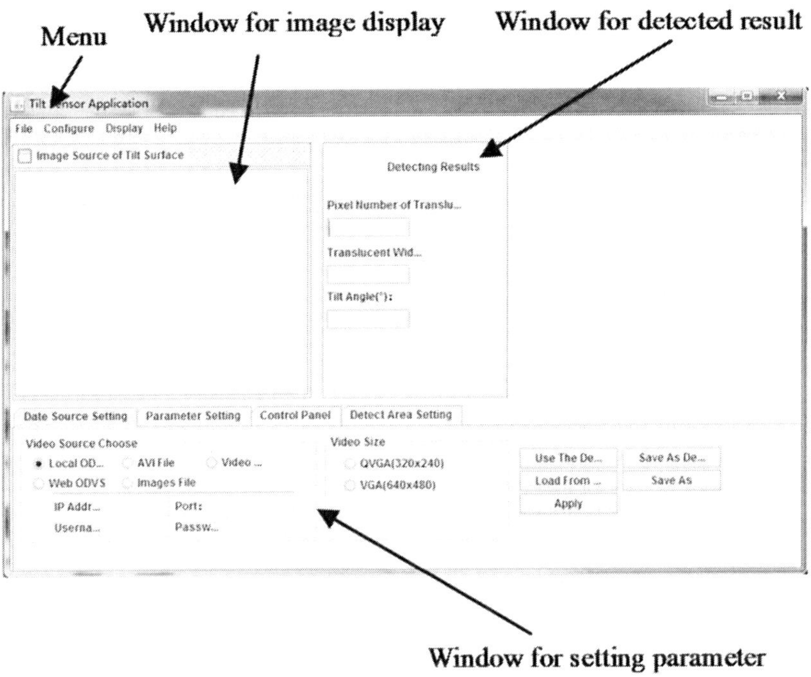

Figure 7. User interface of ODTS application software.

The experimental results show that the detecting precision is relatively high. Moreover, because the detecting algorithm is mainly based on the analysis of the captured image, the detecting range can be very wide (0° ~ 90°). With the different video sources, all the images are processed one by one, thus the ODTS would has good realtime.

The ODTS in this paper includes a camera and an embedded system. It doesn't need the special power supply and the power of the camera and the LED light is offered by the embedded system. When the ODTS doesn't tilt, the embedded system is in dormant state, when it tilts, the mercury switch is connected and the embedded system is in working state (watchdog technology for saving energy). The fully sealed ODTS can work in harsh environment and the working temperature range of the mercury, camera, LED in the tilt sensor is wide. So we can conclude the designed ODTS has simple maintenance, low cost, reliability and high visualization.

Figure 8. The image shape of mercury when ODTS is horizontal.

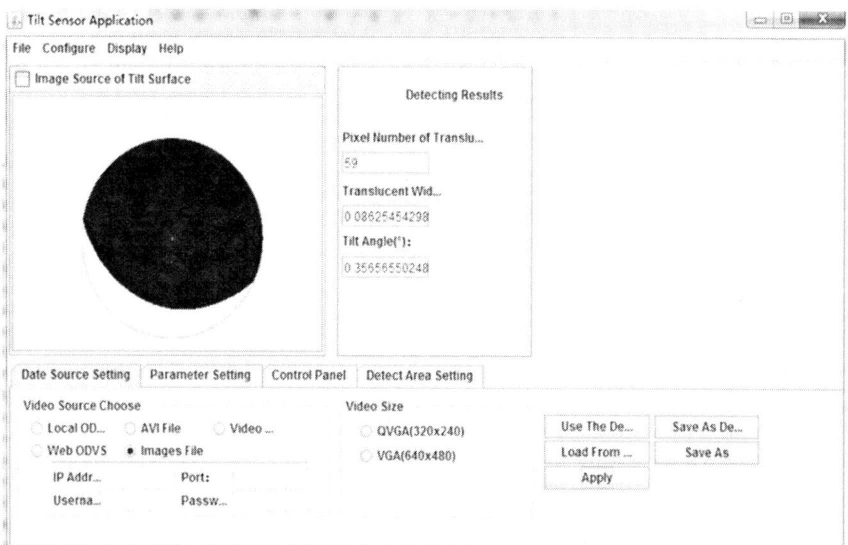

Figure 9. The image shape of mercury when ODTS tilts 7°.

Table 1. The detected results in different artificial tilt angle.

Tilt angle artificially (degree)	7	9	12	15
Pixel number of translucent middle part	59	68	81	92
Middle width of translucent part (mm)	11.9	13.7	16.3	18.5
Detected tilt angle (degree)	6.98	8.63	11.46	14.40
Error (degree)	−0.02	−0.37	−0.54	−0.6
Relative error (%)	−0.28	−4.1	−4.5	−4.0

Now the experiment for prototype is done on the computer. In the future, the embedded system seen as a central controller will be added into the ODTS and gives the signal to start capturing image when the detected surface tilts. At same time, the detected results will be displayed on the interface of the embedded system. The embedded system can also have internet interface to realize remote measurement easily. With remote accessing, we can see the video image of mercury as well as directly get the tilt angle, tilt directional angle and other information at remote computer when the ODTS tilts. Moreover, the embedded system equipped with platform automatic adjustment software would be seen as the upper computer of platform leveling to directly control the lower computer of anchor position servo system and to realize the automatic leveling. Two ways are available to get the access to the measurement data. Users can directly access the system database or also use the API interface provided by the system software. According to the special needs of users, the system can be equipped with wireless communication unit to support various wireless communication network accesses.

5. CONCLUSIONS AND FUTURE WORK

In this paper, we have proposed a novel tilt sensor (OmniDirectional Tilt Sensor) based on machine vision and presented a reliable detecting algorithm. A prototype has been built up and evaluated. The experimental results show that accuracy of 1° and resolution of 0.01° can be achieved. Compared with other tilt sensors, the ODTS has following main advantages: simple maintenance, high precision, wide range, low cost, real-time, reliability and high visualization.

The content of future study will focus on the applications of the ODTS, such as Car alarm electronics, dam safety monitoring, mudslides, landslides and other security monitoring area, offshore platforms, land surveying instrument automatic level school, swing control of ship, ship with chairs, ship bed, mouse, game input devices, civil engineering, water engineering, weapons platforms (tanks, mobile arms control plane), the radar antenna, bridge engineering, agricultural machinery (land leveling, combines, uenae machine and other tilt control.), earthquakes, volcanic detection, vehicle chassis, automatic leveling, automatic welding machines, automatic cutting machine, aviation, aerospace, mobile satellite antenna and other applications that requiring high real-timing. According to the different applications, different ODTSs all based on machine vision would be produced. The study of intelligent video analysis technology with the combination of machine vision to develop series of ODTS for the practical application should be furthered and the measurement precision should be improved further. With the further study, the ODTS that is simple in maintenance, adjustment and calibration, high in accuracy and visualization, wide in measurement range, low in cost, good in remote access, real-time, security and reliability, will be created and applied to important.

REFERENCES

1. C.-T. Wang, E.-H. Yang, J.-G. Zhang, Y.-X. Ma, B. Zhang, "The Omni-Directional Electronic Level Structure and Measuring Analysis Principle," Journal of Scientific Instrument, Vol. 27, No. 2, 2006, pp. 183-185.

2. J.-F. Liu and M.-Y. Li, "The Principle Analytics of Gravity Sensor Level Measuring Instrument," Sensors and Instrumentation, Vol. 24, No. 7, 2008, pp. 159-160.

3. Y. Wu, Y.-H. Song and W.-J. Su, "The Research of Liquid Pendulum Tilt Sensor," Coal Mine and Machinery, No. 6, 2004, pp. 31-32.

4. Y. Lin, L.-H. Pu and F.-X. Zhang, “The Structure Principle of the Omni-Directional Gas Pendulum Tilt Sensor,” Electronic Components and Materials, Vol. 25, No. 5, 2006, pp. 19-22.

5. J.-H. Zhou, J.-Z. Wu and M.-Q. Xu, “The Micro Tilt Sensors of Image-Based Identification,” Automation, Vol. 27, No. 11, 2006, pp. 39-44.

6. B.-O. Guan, H.-Y. Tam and S.-Y. Liu, “TemperatureIndependent Fiber Bragg Grating Tilt Sensor,” IEEE Photonics Technology Letters, Vol. 16, No. 1, January 2004, pp. 244-226.

7. F. Hua, I. Reading and Z. P. Fang, “Novel Opical Sensor for Precise Tilt Angle Measurement,” Control, Automation, Robotics and Vision, 2006.ICARCV’06. 9th International Conference, Singapore, 5-8 December 2006, pp. 1-4.

8. H.-Q. Liu, “Circle Detection Based on Hough Transform,” Master’s Thesis, Northeastern University, Shenyang, 2005, pp. 34-38.

Chapter 2

Monocular Vision Based Boundary Avoidance for Non-Invasive Stray Control System for Cattle: A Conceptual Approach

Adeniran Ishola Oluwaranti*, Seun Ayeni

Department of Computer Science & Engineering,
Obafemi Awolowo University, Ile-Ife, Nigeria

ABSTRACT

Building fences to manage the cattle grazing can be very expensive; cost inefficient. These do not provide dynamic control over the area in which the cattle are grazing. Existing virtual fencing techniques for the control of herds of cattle, based on polygon coordinate definition of boundaries is limited in the area of land mass coverage and dynamism. This work seeks to develop a more robust and an improved monocular vision based boundary avoidance for non-invasive stray control system for cattle, with a view to increase land mass coverage in virtual fencing techniques and dynamism. The monocular vision based depth estimation will be modeled using concept of global Fourier Transform (FT) and local Wavelet Transform (WT) of image structure of scenes (boundaries). The magnitude of the global Fourier Transform gives the dominant orientations and textual patterns of the image; while the local Wavelet Transform gives the dominant spectral features of the image and their spatial distribution. Each scene picture or image is defined by features v, which contain the set of global (FT) and local (WT) statistics of the image. Scenes or boundaries distances are given by estimating the depth D by means of the image features v. Sound cues of intensity equivalent to the magnitude of the depth D are applied to the animal ears as stimuli. This brings about the desired control as animals tend to move away from uncomfortable sounds.

KEYWORDS

Monocular Vision, Control Systems, Global Positioning System, Wireless Sensor Networks, Depth Estimation

1. INTRODUCTION

Intensive monitoring of domestic cattle has many economic and welfare benefits, but to realize these, timely notifications of changes in animal conditions, environmental and physiological events can advance at a rate for which constant monitoring of cattle would be necessary. Free-ranging cattle are controlled by herders. Holding, monitoring or gathering free-ranging cattle require flexible husbandry practices for efficient and effective low stress animal management. Behavioral theory and practical experience indicate that cattle can be taught to respond to auditory cues [1] . Virtual fences techniques for cattle control are achieved by applying aversive stimuli (sound or vibration) to an animal when it approaches a predefined boundary. It is implemented by a small smart collar integrated with a Global Positioning System (GPS) sensor, flash memory, wireless transceivers, and a small CPU; essentially, each node is a small wireless computing device.

The field of wireless sensor networks particularly focuses on applications involving autonomous use of compute, sensing, and wireless communication devices; applicable to animal tracking and monitoring. The work of [2] developed a moving virtual fence algorithm to control cow herd. A smart collar consisting GPS receiver, PDA, WiFi, flash card for WLAN and audio amplifier with speaker was used for each cow in the herd. The animal is given the boundary of a virtual fence in the form of a polygon specified by its coordinates. The location of the animal is tracked against this polygon using the GPS sensor mounted on the collar. When in the neighborhood of a fence, the animal is given a sound stimulus whose volume is proportional to the distance from the boundary, designed to keep the animal within boundaries. In Figure 1 a typical collar is attached around the neck of a cow; the GPS sensors and other sensor nodes are fixed on the collar.

Cattle domain experts have suggested using a library of naturally occurring sounds that are scary to the animals; such as a roaring tiger, a barking dog, a hissing snake; and randomly rotating between the sounds [3] . In most of the existing works on virtual fencing, boundaries of fences are static. Real world herding and control of cattle require dynamic and extensible virtual fences. This work will use monocular vision system for the identification of boundaries, a target recognition subsystem for identification of predefined boundaries to be avoided by cattle. Depth estimation from image structure using the concept of global Fourier transform and local wavelet transform will be used. Sound stimuli generated would be corresponding to the depth estimate.

The monocular vision boundary avoidance for non-invasive stray control system for cattle will require no motion planning algorithm to shift boundaries of virtual fences. It will also require no path planning protocol. This will resolve the inefficiencies in the existing virtual fencing schemes. The rest of this paper is arranged as follows. In Section 2, we present reviews of related works. We describe the overview of a monocular vision based control system for non-invasive stray control of cattle in Section 3. In Section 4, the proposed model formulation is presented while Section 5 offers conclusions and future work.

2. RELATED WORKS

Preliminary research has demonstrated that cows can be controlled and gathered autonomously using recorded audio cues associated with manual gathering [1] . Many works have also been done and are being done on animal tracking and monitoring mostly implemented with wireless sensor networks (WSN) and smart collars.

LynxNet: Wild Animal Monitoring using Sensor Networks presented by [4]; was specifically designed to improve animal monitoring sensor system with low level software node control and communication. The methodology employed involved hardware design prototype of a highly mobile, energy efficient monitoring system that gathers accurate GPS position and multinodal sensor data and disseminates it control. The system of delay tolerant networked nodes to the consumer was used. The mobile nodes initiate all communication, which periodically poll for nearby base stations. The data collected at each node is transferred via radio frequency communication to an end user as shown in Figure 2. The forwarded data is analyzed to determine the behavioral pattern exhibited by the animals being monitored.

The work of [3] presented Wireless Sensor Devices for Animal Tracking and Control is an improvement wireless hardware devices in terms of radio range, solar capabilities, mechanical and electrical robustness and a multi sensor unit for animal tracking. The work carried out experiments with mote hardware using Atmel Atmega128 processor, and the Nordic NRF903 radio chip to prototype and debugs the electronics. The solutions proffered include presentation of a new family (Flock Family) of wireless sensor devices for animal tracking; developed with different engineering tradeoffs to existing devices and to meet the needs of particular pastoral environmental applications.

Figure 1. An animal carrying a sensor node [3] .

The work of [1] presented cows gathering using virtual fencing methodologies. The main objective of the work is to achieve gathering of animals with minimal human interaction using audio cues administered from Direction Virtual Fencing (DVF) equipment. DVF combines a global positioning system (GPS) technology and animal conditioning using sound and mild electrical stimulation; a directional movement of animal using bilateral cues. The draw backs are: cow training protocol required to optimize autonomous gathering has yet to be fully developed; and wireless radio link between base station (in WSN) and the cows used was unpredictable and relatively short. Hence there was high level of equipment failure.

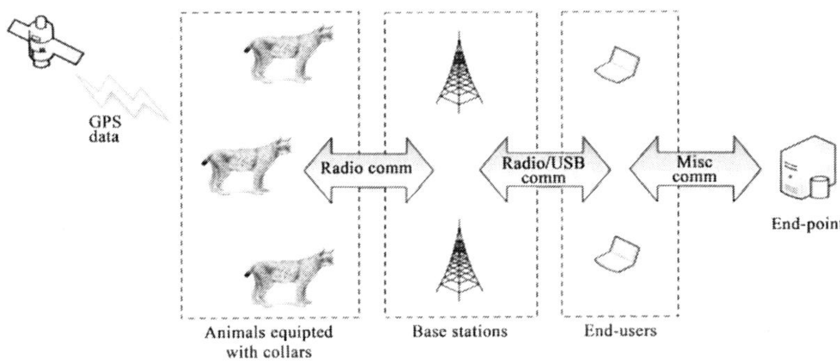

Figure 2. Lynx net system architecture [4] .

Herds of animals such as cattle are complex systems; there are interactions with the environment such as looking for a water source(s) and search for green pasture; necessitating frequent migration from one location to another; with existing polygon coordinate virtual fences techniques limited control can be exerted to effectively move the animals around.

In this work we combine Artificial Intelligence; Networking (WSN) and Control Systems to develop a non- invasive stray control of animal (cattle) to unwanted environments.

3. OVERVIEW OF MONOCULAR BASED MODEL FOR NON-INVASIVE STRAY CONTROL SYSTEM OF CATTLE

Based on the existing work, cattle movement control is achievable with the adaptation of wireless sensor network (WSNs) technological innovations, Global Positioning System (GPS) for animal tracking schemes. This paper proposes a model to enhance the existing work on cattle movement control with the application of wireless sensor network (WSNs); Global Positioning System (GPS); Artificial Intelligence (Image Processing); Monocular Vision System and Target recognition sub-systems.

3.1. Cattle Monitoring with Wireless Sensor Network

Wireless sensor network in cattle monitoring schemes; and precisely in movement control is usually implemented with a sensor mounted collars placed on animals to track individual animal activities and it movement. One notable example of such is “ZebraNet” [5] . Devices mounted on the Zebra routinely exchange data such as GPS position with other devices that fall within their transmission range. If sufficient memory space is available, a user could download historical position data of multiple animals by approaching a single sensor node. The scheme in “ZebraNet” is based on a store and forward approach; the approach may however suffer a considerable delay in packet delivery or loss if the stored data is not frequently updated with the newly measured data.

3.2. Global Position System in Cattle Tracking

Signal that GPS satellites transmit to earth indicate the location of the satellites. A GPS sensor receives the signals and determines the location of animals carrying the sensors by measuring the distance between itself (sensor) and the satellite. There are at least 24 active GPS satellites orbiting around earth at all times. To get a reliable location reading, a typical GPS sensor combines the signals from a number of satellites. Cattle position data and migration pattern acquired, is transmitted node-to-node in the WSN. Data download can then be done from nearest base station for analysis. Over the last decade, many species of terrestrial wildlife have been fitted with GPS collars. Examples of such animals are moose [6] [7] .

GPS tracking collars have been in cooperated in research on the ecology and management of grazing systems using sheep [8] and cattle [9] . By using GPS units in conjunction with Geographic Information Systems (GIS), animal distribution and movement can be evaluated and determined.

3.3. Monocular Vision System for Boundary Avoidance

Generally, obstacles could be avoided effectively with several sensors including infrared, sonar or laser [10] . A direct application found in robotics; an aspect of Artificial Intelligence. Monocular vision system estimate the distance of an obstacle (boundary) from a camera based on the image structures obtain by the camera; known as Depth Estimation. Depth estimation or extraction is the set of techniques and algorithms aiming to obtain a representation of a spatial structure of a scene or boundary. In other terms, to obtain a measure of the distance of ideally, each point of the scene or image structure.

Depth estimation from monocular cues consists of three basic steps. Initially a set of images and their corresponding depth maps are gathered (Arial Depth). Then suitable features are extracted from the images. Based on the features and the ground truth depth maps, learning is done using supervised learning algorithm. The depths of new images are predicted from this learnt algorithm. In a related work [11] used supervised learning to estimate 1-D distances to obstacles, for the application of autonomously driving a remote controlled car.

3.4. Target Recognition Sub-System

Target recognition processes use various methods and algorithms; the sine wave transfer method is used for real-time target recognition, and this signal is used as the process signal in digital signal processing and in analog/digital conversion [12] . Neural network model based and fuzzy logic based algorithms are also being used. The realized target recognition system makes a pixel comparison between the target images pre-recorded in the computer (pre-defined boundaries) and the target image received from the camera.

3.5. Animal Motion Prediction and Control

In the work of [13] ; Data-Driven Identification of Group Dynamics for Motion Prediction and Control; a distributed model structure for representing groups of coupled dynamic agent (Physical bodies) was proposed. The model combines effects from an agent's inertia, interactions between agents, and interactions between each agent and its environment. Global positioning system tracking data were collected in field experiments from group of three (3) cows and a group of ten (10) cows. Least-squares method was used for fitting the model parameters based on measured position data. Based on analysis of model with collected data, hypothesis about cow stray behavioral pattern were proposed.

4. FLOWCHART OF PROPOSED MODEL

In Figure 3 the collar mounted camera takes a capture of the scene (boundary); the captured image is transmitted node-to-node in the WSN. Data (captured image) download can then be done from nearest base station for recognition by the target recognition sub-system. If scene is a recognized boundary; the monocular vision based depth estimator estimates the distance of camera (animal) from the boundary (captured scene). Audio cue of an intensity equivalent to the magnitude of the distance estimated is generated and amplified. The amplified audio cue is applied to the animal (animal left or right ear). Animal responses in terms of animal new or current location; motion direction and speed (velocity) are then tracked by the GPS sensor. The tracked responses determine the next state of the camera.

The general flow chart diagram in Figure 3 shows the steps involved in the entire monocular vision based non-invasive stray control system. At the start of it operation the camera takes a snapshot of a boundary scene. The target recognition subsystem determines if the boundary is a "known boundary" or a "not known boundary". If the boundary is a "known boundary", that is the captured image of the boundary scene matches a pre-recorded image by a similarity of 75% or higher. Then the depth of the image (its distance from the camera) will be estimated using monocular vision. If the captured image does not match any pre-recorded image, then the boundary is "not a known boundary" and the image is simply discarded.

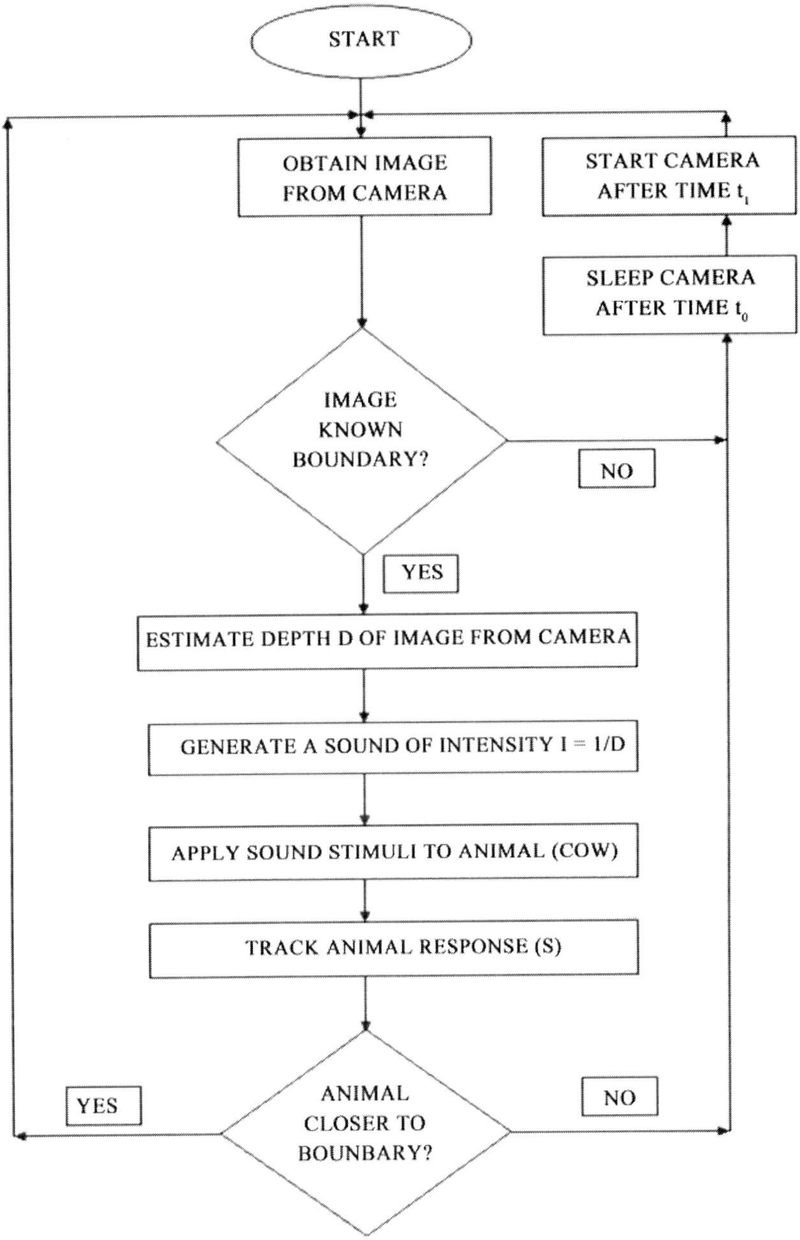

Figure 3. General flow chart diagram of the model.

The sound cue generated is directly proportional to the image estimated distance (D). The sound cue is applied to the animal (cows) ears as a stimulus to give directional and dynamic control to the cow, hence the herds of cattle. It is suggested that pre-recorded scary sounds such as hissing snake or roaring lion be used to control cows away from unwanted environments [14].

4.1. Target Recognition Sub-System

Pre-defined Boundaries: P_B (Images pre-recorded on the Computer system).

Images of Boundaries Received from Camera: P_C

Target Recognition Sub-system compares P_C to P_B

$$\text{Similarity\%} = 100 - \left[\frac{d}{c} \times 100\right] \tag{1}$$

"d" indicates the total number of pixels scanned while "c" indicates the total number of different pixels. If Similarity (%) is 75% or higher match, then scene or boundary is a "known boundary". If similarity (%) is <75%, then scene or boundary is "not a known boundary". "Known boundaries" are to be avoided by animals (cattle); "not a known boundary" may or may not be avoided by animals (cattle). Pixels based calculation of middle point of object (image). The formula as follows [15] :

$$\text{middle point}\left(x_g, y_g\right)$$

$$x_g = \frac{\int x \mathrm{d}L}{\int \mathrm{d}L} \tag{2}$$

$$y_g = \frac{\int y \mathrm{d}L}{\int \mathrm{d}L} \tag{3}$$

The coordinates of the middle point are found as an ordered series obtained by calculating the Euclidean distance (E) between the mid point (x_g, y_g) and each border pixel (x_i, y_i).

$$E = \sqrt{(x_i - x_1)^2 + (y_i - y_1)^2} \tag{4}$$

where $i = 1,2,3,4,\cdots,N$; N is the number of corner pixels.

4.2. Monocular Depth Estimation Model

The following is a description of image structure based on the textual patterns present in the image and their spatial arrangement [16] [17] .

First level description: Magnitude of the global Fourier transform of the image. The discrete Fourier transform (FT) of an image is defined as [10] .

$$I(f) = \sum_{x=0}^{N-1}\sum_{y=0}^{N-1} i(\times)h(\times)\mathrm{e}^{-j2\pi\langle f,x\rangle} \tag{5}$$

where i(x) is the intensity distribution of the image along the spatial variables $x = (x, y); j = \sqrt{-1}$; and the spatial frequency variables are defined by $f = (f_x, f_y) \in [-0.5, 0.5] \times [-0.5, 0.5]$ (units are in cycles per pixel); and h(x) is a circular window that reduces boundary effects.

Second level description: Wavelet transforms defined in [10] as:

$$I(x,k) = \sum_{x'} i(x')h_k(x - x') \tag{6}$$

Different choices of the functions h_k provide different representations for instance as in [15] . One of the most common functions is complex Gabor Wavelets:

$$h_k(x) = \mathrm{e}^{-\pi\|x\|^2/\sigma_k^2}\mathrm{e}^{-2\pi j\langle f_k, x\rangle} \tag{7}$$

In such a representation, I(x, k) is the output at the location ´ of a complex Gabor filter turned to the spatial frequency defined by f_x. Our objective is to estimate the depth D of the scene (boundary) by means of the image features, v [10] .

$$D/V = \int_{-\infty}^{\infty} D f_{D/V}\left(\frac{D}{V}\right) \mathrm{d}D \tag{8}$$

Where

$$f_{D/V}(D/V) = F_{D/V}(D \cdot V)/f_v(v) \tag{9}$$

And

$$f_v(v) = \int f_{D,V}(D,V)\,\mathrm{d}D \tag{10}$$

4.3. Animal Motion Prediction Model

Given a group of m agents, the model structure for an individual agent $i \in \{1, \cdots, m\}$ can be written in state-space, difference equation form as in [13]

$$x_i^{\tau+1} = \begin{bmatrix} 1 & 0 & \Delta t & 0 \\ 0 & 1 & 0 & \Delta t \\ 0 & 0 & a_i & 0 \\ 0 & 0 & 0 & a_i \end{bmatrix} x_i^\tau + \begin{bmatrix} 0 & 0 \\ 0 & 0 \\ 1 & 0 \\ 0 & 1 \end{bmatrix} \times \left[\sum_{j=1, j \neq 1}^{m} f_{ij}\left(P_i^\tau, P_j^\tau\right) + g_i\left(P_i^\tau\right) + w_i^\tau \right] \tag{11}$$

Agent i's state $x_i^\tau = \left[e_i^\tau n_i^\tau u_i^\tau v_i^\tau\right]^{\mathrm{T}}$ consists of its east position, north position, eastern component of velocity, and northern component of velocity after the τth iteration, and its position if given by $P_i^\tau = \left[e_i^\tau n_i^\tau\right]^{\mathrm{T}}$. The time step Δt is given by $t^{\tau+1} - t^\tau$, constant for all τ. The term a_i represents damping, $a_i = 1$ for zero damping and $|a_i| < 1$ for stable systems. The function $f_{ij}\left(P_i^\tau, P_j^\tau\right)$ determines the coupling force applied by agent j to agent i. The function $g_i(P_i)$ represents the force applied by the environment to the agent at point P_i^r . W_i^τ is a Zen mean, stationary, Gaussian white noise. Process uncorrelated with $P_j \forall_j$ used to model the unpredictable decision-motive processes of agent i.

4.4. Algorithm for the Camera Sequence or Mode of Operations

Step 1: Start (Camera)

Step 2: Snapshot (Camera)

Step 3: Sleep (Camera) For time $t = t_1$ (min)

Step 4: Compute F(v) = 1/V

Step 5: Set time $t_2 \leftarrow$ F(v)

Step 6: If time $t > t_2$ (min)

Step 7: Then Sleep (Camera) for time $t = t_2$

Step 8: else Sleep (Camera) for time $t = t_1$

Step 9: wake (Camera)

Step 10: End (Camera)

The camera sleeps or delay it operation for time t (min) before "waking-up" again to capture images. The variable t can assume or be assigned two values t_1 and t_2. Time t_1 (min) is a predefined camera delay time determined by the user. The value of t_1 can be RESET; after which it is assigned to t. The values of t_1 can be SET to 5 minutes or 10 minute as predetermined by the user. Time t_2 is modeled as a function of animal (cow) velocity V; given as F(v) = 1/V. Time t_2(min) is assigned the value of F(v) each time it is computed. Then camera delay time t is assigned the value of t_2.

4.5. Simulation of Camera Sequence or Mode of Operations

Figure 4 shows the simulation of the relationship (inverse proportionality) between camera delay time t_2 (min) and animal velocity V (m/s). This indicates that when animal velocity towards a boundary is high, the camera delay time t_2 is small. Time t_2 (min) reduces as animal velocity V (m/s) increases.

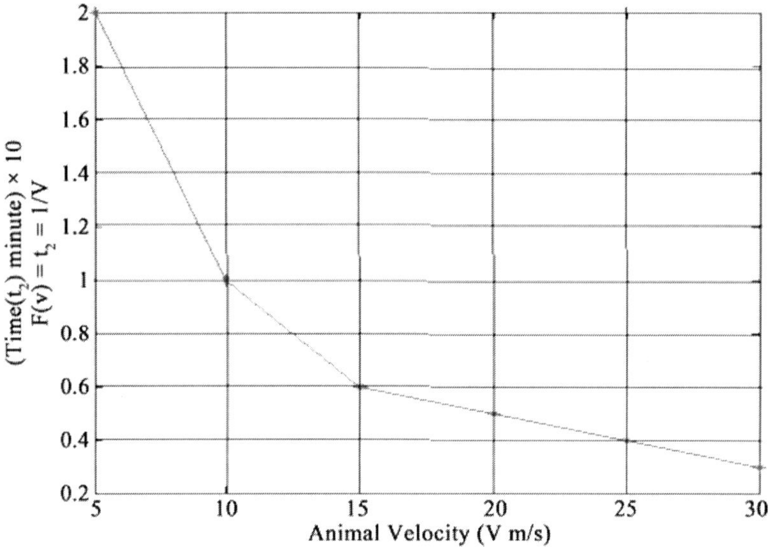

Figure 4. Graph of camera delay time t_2 (min) against animal velocity V (m/s).

5. CONCLUSION

In this paper, we proposed a model for an improved virtual fencing stray control system for cattle. The proposed model expressed as an algorithm, uses a target recognition sub-system to determine the boundary to be avoided; monocular vision based estimation of boundary distances and sound cues aversive stimuli to control or enforce animal movement away from protected environments. This is a substantial and an improved approach over existing animal control techniques.

FURTHER WORKS

Future works will be to evaluate the depth error perception and the stability of the non-invasive stray control system. This could be done using an appropriate simulation tool and the performance metrics of interest. The results of the simulation could be benchmarked with the existing schemes in order to determine it efficiency.

REFERENCES

1. Dean, M., Doniec, M., Detweiler, C., Vesilecu, L., Rus, D., Nolen, B. and Libeau, R. (2008) Gathering Cows Using Virtual Fencing Methodologies. Computer Science and Artificial Intelligence Laboratory, Cambridge.

2. Bluler, Z., Corke, P., Peterson, R. and Rus, D. (2004) Networked Cow: Virtual Fences for Controlling Cows. Proceedings of IEEE Conference on Robotics and Automation.

3. Sikka, P., Corke, P. and Overs, L. (2004) Wireless Sensor Devices for Animal Tracking and Control. Proceedings of the 29th Annual IEEE International Conference on Local Computer Networks (LCN'04), 16-18 November 2004, 446-454.

4. Zviedris, R., Elst, A., Strazdus, G., Medrus, A. and Selavo, L. (2010) LynxNet: Wild Animal Monitoring Using Sensor Networks. http://www.springlink.com/content/573467t6gw893146/

5. Juang, P., Oki, H. and Wang, J. (2002) Energy-Efficient Computing for Wildlife Tracking: Design Tradeoffs and Early Experiences with ZebraNet. Proceedings of the 10th International Conference on Architectural Support for Programming Language and Operating Systems (ASPLOS-X).

6. Moen, R. and Colien, Y. (1996) Interpreting Behavior from Activity Counters in GPS Collars on Moose. Alces, 32, 101-108.

7. Rodgers, A. and Anson, P. (1994) Animal-Borne GPS: Tracking the Habitat. GPS World, 5, 20-32.

8. Rutter, M., Beresford, A. and Robert, G. (1997) Use of GPS to Identify the Grazing Areas of Hill Sheep. Computer and Electronics in Agriculture, 17, 177-188.

9. Turner, L., Ubal, M., Larson, B. and Shearer, A. (2000) Monitoring Cattle Activity and Pasture Use with GPS and GIS. Canadian Journal of Animal Science, 80, 405-413.

10. Torralba, A. and Oliva, A. (2002) Depth Estimation from Image Structure. IEEE Transactions on Pattern Analysis and Machine Intelligence, 24, 1226-1238.

11. Saxena, A., Chung, S. and Andrew, Y. (2006) Learning Depth from Single Monocular Images. In NIRS 2006. 1, 2, 4.

12. Yagimli, M. and Varol, H. (2009) A GPS-Based System Design for the Recognition and Tracking of Moving Targets. 4th International Conference on Recent Advances in Space Technologies, Istanbul, 11-13 June 2009, 6-12.

13. Schwager, M., Detweiler, C., Vasilescu, L., Anderson, D. and Rus, D. (2008) Data-Driven Identification of Group Dynamics for Motion Prediction and Control. Journal of Field Robotics, 25, 305-324.

14. Dean, M. (2000) The Cyber Cow Whisperer and His Virtual Fence. Agricultural Research. http://www.nps.ars.usda.gov

15. Simonceli, E. and Freeman, W. (1995) The Steerable Pyramid a Flexible Architecture for Multi-Scale Derivative Computation. IEEE Transactions on Image Processing.

16. Oliva, A., Torralba, A., Guerin-Dugue, A. and Herault, J. (1999) Global Semantic Classification Using Power Spectrum Template. Proceedings of the Challenge of Image Retrieval.

17. Oliva, A. and Torralba, A. (2001) Modeling the Shape of the Scene: A Holistic Representation of the Spatial Envelop. International Journal of Computer Vision, 42, 145-175.

Chapter 3

A Vision-Based Sensor for Noncontact Structural Displacement Measurement

Dongming Feng *, Maria Q. Feng, Ekin Ozer and Yoshio Fukuda

Department of Civil Engineering and Engineering Mechanics, Columbia University, New York, NY 10027, USA

ABSTRACT

Conventional displacement sensors have limitations in practical applications. This paper develops a vision sensor system for remote measurement of structural displacements. An advanced template matching algorithm, referred to as the upsampled cross correlation, is adopted and further developed into a software package for real-time displacement extraction from video images. By simply adjusting the upsampling factor, better subpixel resolution can be easily achieved to improve the measurement accuracy. The performance of the vision sensor is first evaluated through a laboratory shaking table test of a frame structure, in which the displacements at all the floors are measured by using one camera to track either high-contrast artificial targets or low-contrast natural targets on the structural surface such as bolts and nuts. Satisfactory agreements are observed between the displacements measured by the single camera and those measured by high-performance laser displacement sensors. Then field tests are carried out on a railway bridge and a pedestrian bridge, through which the accuracy of the vision sensor in both time and frequency domains is further confirmed in realistic field environments. Significant advantages of the noncontact vision sensor include its low cost, ease of operation, and flexibility to extract structural displacement at any point from a single measurement.

KEYWORDS

vision sensor; displacement; template matching; upsampled cross correlation; subpixel resolution; civil engineering structures

1. INTRODUCTION

Civil engineering structures including buildings and bridges are exposed to various external loads such as traffic, gust and earthquake during the operational lifetime. Monitoring structural static and dynamic displacements can provide quantitative information for both structural safety evaluations and maintenance purposes. Such practice, however, is highly expensive to operate, mainly due to cumbersome, time-consuming, and expensive installation of sensors and their data acquisition systems. Sensors currently available for measuring structural displacements can be classified as contact-type (e.g., Linear Variable Differential Transformer (LVDT)) and noncontact-type (e.g., GPS, laser vibrometer and radar interferometry system) sensors [1,2,3,4,5,6,7,8]. The contact sensor requires the access to the object structure to install the sensor and physically connect it to a stationary reference point, which is often difficult or even impossible to locate. The GPS sensors are easier to install but the measurement accuracy is limited, usually with errors between 5 and 10 mm [1,2,3,4]. The noncontact laser vibrometer is generally accurate. But the limited measurement distance prevents its applications in monitoring civil engineering structures because longer distance measurement requires the use of a high-intensity laser beam that would endanger human health [4,6]. The interferometric radar system allows remote measurements with a good resolution. However, it requires reflecting surfaces mounted on the structure [5].

To cope with these problems, noncontact vision-based displacement measurement systems have been developed recently, which are primarily enabled by the template matching/registration techniques [1,2,3,9,10, 11,12,13,14,15,16,17,18,19]. For example, Busca *et al.* [12] developed a vision-based displacement sensor system using three template matching algorithms, namely, pattern matching, edge detection and digital image correlation (DIC). The vision sensor was used to measure the vertical displacement of a railway bridge by tracking high-contrast target panels fixed to the bridge. Song *et al.* [11] measured the displacement of a cantilever beam from a vision sensor by extracting markers using subpixel Hough transforms from video images. Kim *et al.* [9] proposed a vision-based monitoring system using DIC to evaluate the cable tensile forces of a cable-stayed bridge. Ribeiro *et al.* [3] measured the dynamic displacement of a railway bridge utilizing the RANdom SAmple Consensus (RANSAC) algorithm. On the basis of a robust orientation code matching (OCM) algorithm, the authors developed a vision sensor system for real-time displacement measurement by tracking natural targets on the structural surface, which eliminates the requirement for physical access to structures to install artificial target panels [2,20].

In practice, one major concern for the vision sensor system is the measurement accuracy. Template matching technique usually gives displacement with integer-pixel resolution since the minimal unit in a video image is one pixel. Although in many applications the pixel-level accuracy is adequate, it is often far from the required in case of small structural vibrations. To improve the measurement precision, incorporating the subpixel registration into the template matching algorithm is regarded as the best choice. The interpolation technique is most commonly used subpixel approach, examples of which includes intensity interpolation, correlation coefficient curve-fitting or interpolation, phase correlation interpolation and the geometric methods [21,22,23]. Subpixel registration can also be formulated as an optimization problem and solved through heuristic algorithms such as genetic algorithm, artificial neural network algorithm, and particle swarm optimization, *etc.* [24,25]. There are also other subpixel techniques that are based on Newton-Raphson method [26] and gradient-based methods [27].

As an emerging sensor technique, there is a need to thoroughly investigate the noncontact vison sensor by employing different template matching as well as subpixel algorithms, and experimentally evaluating its performance through tests on various structures. In this study, a novel vision sensor is developed based on an advanced subpixel template matching technique, *i.e.*, the upsampled cross correlation (UCC), which is developed into a software package for real-time displacement extraction. A series of laboratory and field tests are carried out to evaluate its performance.

The paper is organized as follows: in Section 2, the estimation of scaling factor is discussed, and the vision sensor system including the hardware and the theoretical background of the software is introduced; Section 3 evaluates its performance through a laboratory shaking table test of a small-scale frame structure; Section 4 presents two field tests results of a railway bridge and a pedestrian bridge, respectively; Section 5 concludes this study.

2. PROPOSED VISION SENSOR SYSTEM

The underlying principle of the vision sensor for displacement measurement is the template matching technique. Figure 1shows the basic procedure of the vision sensor implementation. In the implementation, an initial area to be tracked is defined as a template in the first image of a sequence of video frames. The template can be located in the successive images using the template matching technique. To reduce computational time, the searching area could be confined to a predefined region of interest (ROI) near the template's location in the previous image.

2.1. Scaling Factor Determination

In order to obtain structural displacements from the captured video images, the establishment of the relationship between the pixel coordinate and the physical coordinate is required (e.g., with units of mm/pixel). As shown in Figure 2a, when the image plane is parallel to the object surface, the scaling factor in the translational direction (*x* axis) can be determined by:

$$SF = \frac{d_{known}}{I_{known}} \quad \text{or} \quad SF = \frac{d_{known}}{d^{i}_{known}} d_{pixel} = \frac{D}{f} d_{pixel} \tag{1}$$

where dknown is the known physical length on the object surface, diknown and Iknown are the corresponding physical length and pixel length at the image plane respectively with diknown=Iknowndpixel, dpixel is the pixel size (e.g., in μm/pixel), *D* is the distance between the camera and the object, and *f* is the focal length.

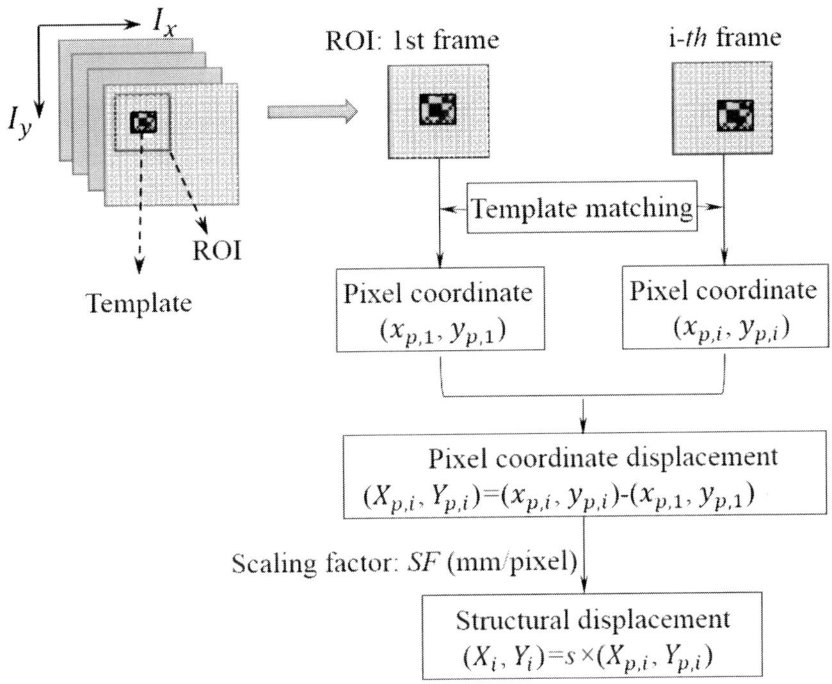

Figure 1. Procedure of vision sensor implementation.

Thus the scaling factor can be obtained from one of the two methods: (1) be estimated from the known physical dimension on the object surface and its corresponding image dimension in pixels (*i.e.*, dknown and Iknown); (2) be estimated based on the intrinsic parameters of the camera as well as the extrinsic parameters between the camera and the object structure (*i.e.*, *D*, *f*and dpixel).

However, the prerequisite of Equation (1) is the perpendicularity of the camera's optical axis to the object surface. Thus all points on the object surface have equal depth of fields [8,28]. Such a requirement would impose some difficulties in the practical implementations because small magnitude of camera misalignment angle can be unnoticed during the experiment setup especially when the object distance from the camera is relatively large. Moreover, in outdoor field tests, it is sometimes unavoidable to tilt the camera optical axis by a small angle in order to track the measured object surface.

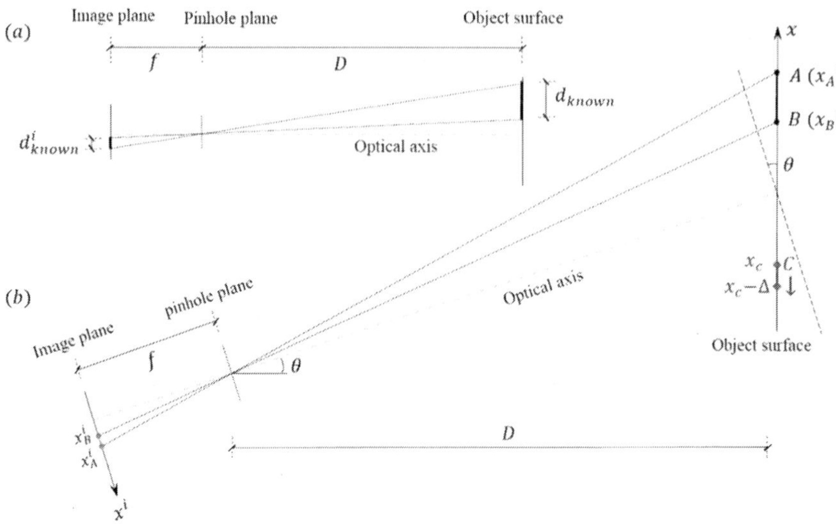

Figure 2. Scaling factor determination: (**a**) optical axis perpendicular to object surface; (**b**) optical axis non-perpendicular to object surface.

Figure 2b shows a schematic when the camera optical axis is tilted about the normal directions of the object surface by an angle θ. Assume line *AB* is known dimension on the object. xA and xB are the coordinates of the two points, and IiA and IiBare the corresponding pixel coordinates at the image plane. The scaling factor can be estimated by:

$$SF_1 = \frac{x_A - x_B}{I^i_A - I^i_B} \quad (2)$$

From the triangular geometry, xA and xB can be expressed as:

$$x_A = \frac{Dx^i_A}{f\cos^2\theta - x^i_A\cos\theta\sin\theta}, \quad x_B = \frac{Dx^i_B}{f\cos^2\theta - x^i_B\cos\theta\sin\theta} \quad (3)$$

where $x^i_A = I^i_A d_{pixel}$ and $x^i_B = I^i_B d_{pixel}$ are the coordinates at the image plane. When θ is small $(\sin\theta \approx 0)$, and $x^i_A \ll f$ and $x^i_B \ll f$, the scaling factor in Equation (2) can be further estimated and simplified in terms of the intrinsic camera parameters and the extrinsic parameters between the camera and the object structure:

$$SF_2 = \frac{1}{I_A^i - I_B^i}\left(\frac{Dx_A^i}{f\cos^2\theta - x_A^i\cos\theta\sin\theta} - \frac{Dx_B^i}{f\cos^2\theta - x_B^i\cos\theta\sin\theta}\right) \approx \frac{D}{f\cos^2\theta} d_{pixel} \quad (4)$$

For example, if point *C* in Figure 2b has a small translation Δ along the *x* axis at the object surface, the "true displacement" is:

$$\Delta = x_C - (x_C - \Delta) = \frac{Dx_C^i}{f\cos^2\theta - x_C^i\cos\theta\sin\theta} - \frac{Dx_C^{i,\Delta}}{f\cos^2\theta - x_C^{i,\Delta}\cos\theta\sin\theta} \quad (5)$$

where $x_C^i = I_C^i d_{pixel}$ and $x_C^{i,\Delta} = I_C^{i,\Delta} d_{pixel}$ are the coordinates of point *C* before and after translation at the image plane.

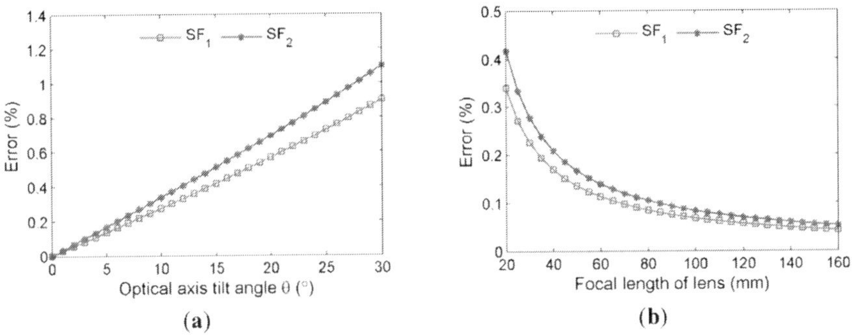

Figure 3. Error resulting from camera non-perpendicularity: (**a**) Effect of optical axis tilt angle (*f* = 50 mm); (**b**) Effect of focal length of lens (ϑ = 3°).

From the scaling factors SF1 in Equation (2) and SF2 in Equation (4), the "measured displacement" can be estimated $\tilde{\Delta}_1 = (I_C^i - I_C^{i,\Delta})\mathrm{SF}_1$ or $\tilde{\Delta}_2 = (I_C^i - I_C^{i,\Delta})\mathrm{SF}_2$ SF2. In order to quantify the error resulting from camera non-perpendicularity, numerical studies are conducted. The measurement errors from the two scaling factors can be defined as: Error=||Δ˜1−Δ||/Δ × 100% and Error=||Δ˜2−Δ||/Δ × 100%. The adopted parameters are: camera with 640 × 512 pixel resolution, dpixel=4.8μm, IiA=200 and IiB=160, *D* = 10 m. Point *C* has a 1 pixel translation in the image plane from IiC=100 to Ii,ΔC=99. The effects of the optical axis tilt angle and lens focal length are investigated by considering a variable range and the results are shown in Figure 3. It can be seen that the error increases as the tilt angle increases and the error is inversely related to the focal length. In sum, it could be concluded that in most practical applications the measurement errors from small optical tilt angles are acceptable. Although this study is based on the 1D (*x* axis) in-plane translation, the conclusions can be equally extended to the 2D in-plane translation.

It is also found from the numerical study that for a fixed camera setup, the measurement error from scaling factor SF1would decrease when the measurement point *C* gets closer to the known dimension *AB*. Especially, the error is minimized when the measurement point is located within the region of known dimension. For scaling factor SF2 in Equation (4), errors would further arise from the uncertainties in the tilt angle estimation, camera distance measurement and focal length readings from the adjustable-focal-length lens.

In the laboratory and field tests of this study, scaling factor SF_1 is adopted, which is obtained from Equation (2) based on the known physical dimension on the object surface (e.g., the size of artificial target panels or the size of the nuts and rivets known from the design drawings) and the corresponding image dimension in pixels. It is noted that camera calibration according to Zhang method [29] would reduce the effect of lens distortion [19,30], which is however not carried out in this study.

2.2. Hardware of the Vision Sensor System

As tabulated in Table 1, the proposed vision sensor system simply consists of a video camera, a zoom lens and a notebook laptop. During the test, the camera equipped with the lens is fixed on a tripod and placed at a remote location away from the structure. The camera is connected to a laptop installed with the real-time image-processing software. It is noteworthy that setting up the vision sensor, including focusing the lens on the targets, takes only a few minutes.

Table 1. Technical specifications of the proposed vision sensor system.

Component	Model	Technical Specifications
Video camera	 Point Grey/FL3-U3-13Y3M-C	Maximum resolution: 1280 × 1024 Frame rate: 150 FPS Chroma: Mono Sensor type: CMOS Pixel size: 4.8 μm Lens mount: C-mount Interface: USB3.0
Optical lens	 Kowa/LMVZ990 IR	Focal length: 9 to 90 mm Maximum Aperture: F1.8 Mount: C-mount
Laptop computer	 Sony /PCG-41216L	Intel(R) Core(TM) i7-2620M CPU @ 2.70 GHz 8192 RAM 250 HDD 14.1" Screen
Tripod and Accessories	Tripod, USB3.0 type-A to micro-B cable, *etc.*	

2.3. Upsampled Cross Correlation for Template Matching

In this study, the vision sensor is developed based on UCC, a subpixel template matching method proposed by Guizar-Sicairos *et al.* [31]. Consider a pair of images f(x,y) and t(x,y) with identical dimensions M×N, among which, t(x,y) has a relative translation from the reference image f(x,y). The cross correlation between f(x,y) and t(x,y) by means of Fourier transform can be defined as:

$$R_{FT}(x_0,y_0)=\sum_{x,y} f(x,y)t^*(x-x_0,y-y_0)=\sum_{u,v} F(u,v)T^*(u,v)\exp\left[i2\pi\left(\frac{ux_0}{M}+\frac{uy_0}{N}\right)\right] \quad (6)$$

where the summations are taken over all image points (x,y); (x0,y0) is an amount of coordinate shift; "*" denotes complex conjugation; F(u,v) and T∗(u,v) represent the discrete Fourier transform (DFT) of their lowercase counterparts, for example

$$F(u,v)=\sum_{x,y}\frac{f(x,y)}{\sqrt{MN}}\exp\left[-i2\pi\left(\frac{ux}{M}+\frac{uy}{N}\right)\right] \quad (7)$$

From Equation (6), an initial displacement estimation with pixel-level resolution can be easily acquired by locating the peak of RFT. Subsequently, cross correlation based on a time-efficient matrix-multiplication discrete Fourier transform (DFT) is performed in a neighborhood around the initial peak to achieve a subpixel resolution.

Figure 4 shows the flowchart of the vision sensor based on the subpixel UCC, described as follows:

Step I: Pixel-level rough search. Compute the cross correlation between the image to register and the reference image by means of Fourier transform, and the initial displacement can be estimated from the correlation peak;

Step II: Subpixel fine search. Compute the cross correlation in a 1.5 × 1.5 pixel neighborhood around the initial estimate by an upsampling factor of κ. Thus a subpixel resolution within 1/κ of a pixel is achieved by searching the peak in this (1.5κ,1.5κ) neighborhood. For example, by setting κ=10, a 0.1 subpixel accuracy can be achieved.

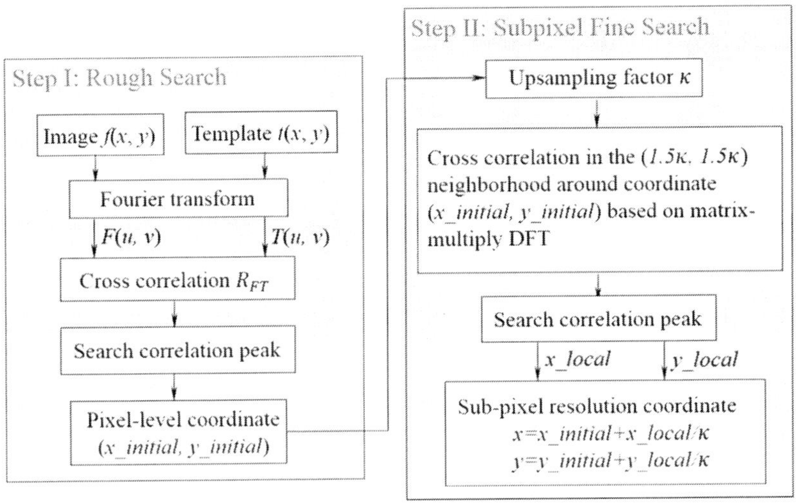

Figure 4. Flowchart of the upsampled cross correlation (UCC) implementation.

In Step II, instead of computing a zero-padded FFT, a matrix-multiplication DFT operation is implemented by the product of three matrices with dimensions (1.5κ,N), (N,M) and (M,1.5κ). The algorithm complexity for this upsampling subpixel search is O(MNκ), while complexity of conventional FFT upsampled by zero-padding F(u,v)T∗(u,v) is O(MNκ[log2(κM) +κlog2(κN)]). The substantial improvement dramatically reduces computational time and memory requirement without sacrificing accuracy, making possible of real-time displacement measurement.

A real-time video-processing software is developed based on UCC. The programming environment for the software package is Visual Studio 2010 using C++ language. During measurement, the FlyCapture Software Development Kit (SDK) by Point Grey Research is used to capture video images from Point Grey USB 3.0 cameras using the same application programming interface (API) under 32- or 64-bit Windows 7/8 operating system. Then the frame-by-frame image are processed by the UCC algorithm and displayed on the screen using DirectShow library. Meanwhile, the measured displacement history would be shown on the screen in real time and saved to the computer. The online measurement avoids the time-consuming and memory-intensive task of saving huge video files. However, a tradeoff among measurement points, video resolution, maximum frame rate per second and template sizes is necessary. On the other hand, the developed software can also be used for post-processing the recorded video files, which enables the flexibility to extract structural displacements at more points from a single recording.

3. SHAKING TABLE TEST OF A FRAME STRUCTURE

The performance of the proposed vision sensor is first evaluated through a shaking table test of a scaled three-story frame structure in the Carleton Laboratory at Columbia University, as shown in Figure 5. The aluminum frame structure is bolt-connected for all the column-floor connections.

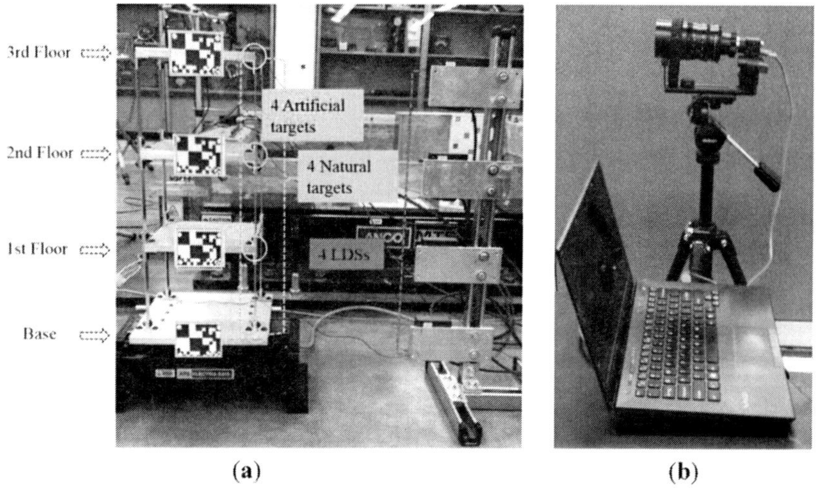

Figure 5. Laboratory test: (**a**) Shaking table test setup; (**b**) Vision sensor system setup.

3.1. Shaking Table Test Setup

During the testing, the shaking table (Model# APS113 by APS Dynamics Inc.: San Juan Capistrano, CA, USA) is driven by white noise signals. Four predesigned black and white artificial targets (99 mm × 75 mm) are mounted on the structure for motion tracking. Meanwhile, four bolt connections are used to study the performance of the vision sensor to track natural targets on the structure. As references, the displacements are also measured by four high-accuracy laser displacement sensors or LDSs (Model#LK-G407 by KEYENCE), which are installed between each floor of the frame model and stationary reference points.

The visions senor system is placed 8 m away from the shaking table. During the measurement, video images captured by the camera are digitized into 640 × 512 pixel images in 8 bit grey scales and streamed into the computer through an USB 3.0 cable. Before testing, the processing time for each video frame by the developed software should be obtained to determine the maximum frame rate. In this test, four small areas from the artificial targets and four bolt-connection areas at all floors are simultaneously registered as templates. It is observed that a total time of 5.6 ms is needed for each video frame, including the reading and preparing, template matching and image displaying time. Thus, real-time displacement time histories at eight measurement points can be simultaneously measured with a sampling rate of 150 fps.

3.2. Subpixel Resolution Performance

Pixel-level template matching may result in unacceptable measurement errors if the displacement to be measured has same order of magnitude as the scaling factor. In this case, the subpixel technique should be adopted to make template matching fall at a fractional pixel location. To better understand how the subpixel technique improves the measurement precision, displacements extracted from video images by tracking the artificial target on the base floor are used as a demonstration. Four subpixel levels, namely, levels of one integer pixel, 0.5 pixel, 0.2 pixel and 0.05 pixel are chosen, with the corresponding resolutions tabulated in Table 2. Recall again, a desired subpixel resolution can be easily achieved by simply adjusting the upsampling factor κ. In this testing, the scaling factor is 1.338 mm/pixel, providing ±0.669 mm resolution.

As shown in Figure 6a, for the integer-pixel resolution (1.338 mm), the displacement errors between the vision sensor and LDS can be observed clearly. On the other hand, after employing different levels of subpixel analysis, the displacement by the vision sensor agrees better with that by LDS as the resolution improves (with NRMSE errors (by Equation (8)) of 6.41%, 3.80%, 1.73%, and 1.35% respectively for the four zoom-in segments in Figure 6).

Table 2. Different levels of subpixel resolution.

Subpixel (pixel)	1	0.5	0.2	0.05
Resolution (mm)	±0.669	±0.335	±0.134	±0.034

It is noted that in the ideal case where video images have no distortion or noise, larger upsampling factor κ would yield smaller error. However, the subpixel accuracies reported in many studies vary within orders of magnitude from 0.5 to 0.01 pixel [21,32], as images may be contaminated by various external environmental noises and system noises arising from the electronics of the imaging digitizer. For the following tests in this study, $\kappa=20$ is selected.

3.3. Measurement Evaluation by Tracking both Artificial and Natural Targets

To evaluate the performance of the vision sensor, displacements are measured by tracking both the black and white artificial targets and natural targets (*i.e.*, bolt connections) and compared with those by four LDSs. Here, the measurements are termed as Vision (artificial target), Vision (natural target) and laser displacement sensor (LDS), respectively.

Firstly, the four artificial targets in Figure 5 are used as the tracking target for the vision sensor. Figure 7 shows the comparison of displacements by Vision (artificial target) and LDS. Excellent agreements can be observed. In Figure 8, the plotted displacements of 1st, 2nd, and 3rd floor are relative to the base-floor displacement in Figure 7a, and only enlarged time segments between 2 s and 4 s are shown for better illustration. As shown in Figure 8a, small discrepancies are observed in the 1st floor relative displacements by the vision sensor and LDS. However, considering the small vibration amplitude (smaller than 1 mm), the errors are acceptable. As can be seen from the relative displacements of 2nd and 3rd floor in Figure 8b,c, the discrepancies are reduced as the vibration amplitude increases. And it is expected that the errors would further decrease as the amplitude of the relative vibration increases.

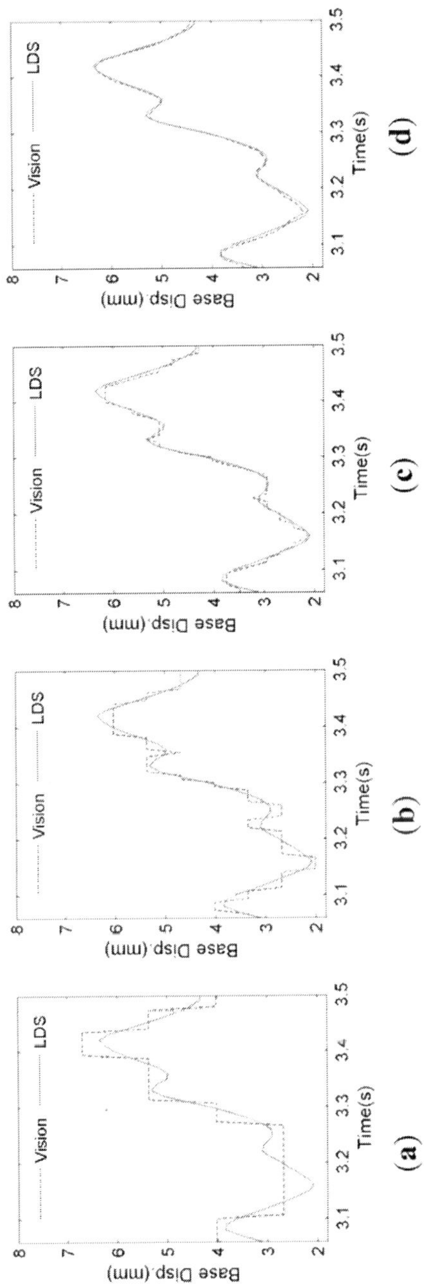

Figure 6. Subpixel resolution evaluation (**a**) Resolution: ±1.338 mm; (**b**) Resolution: ±0.669 mm; (**c**) Resolution: ±0.268 mm; (**d**) Resolution: ±0.067 mm.

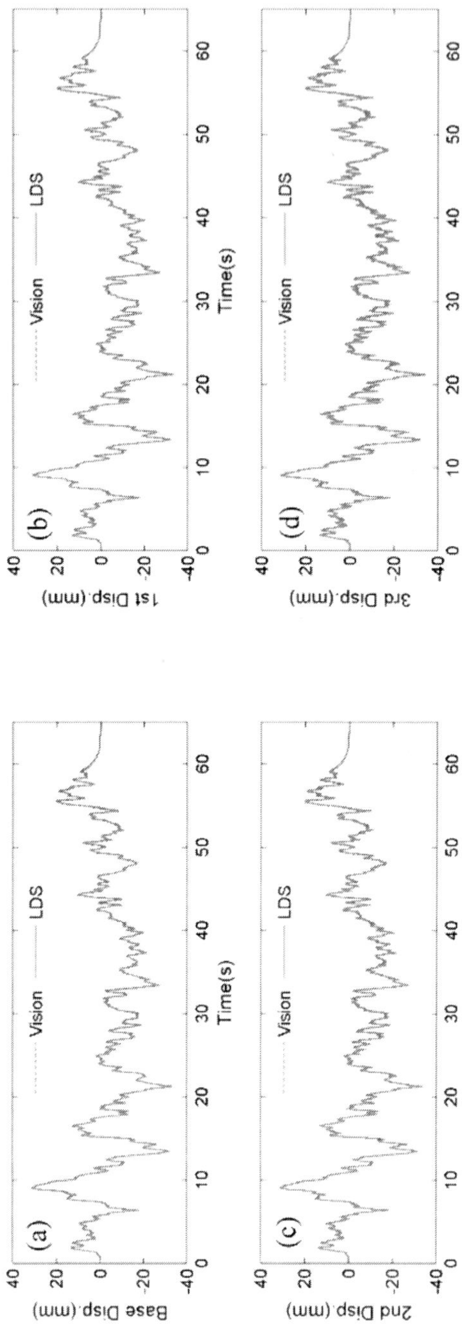

Figure 7. Displacement comparisons between Vision (artificial target) and LDS: (**a**) Base floor; (**b**) 1st floor; (**c**) 2nd floor; (**d**) 3rd floor.

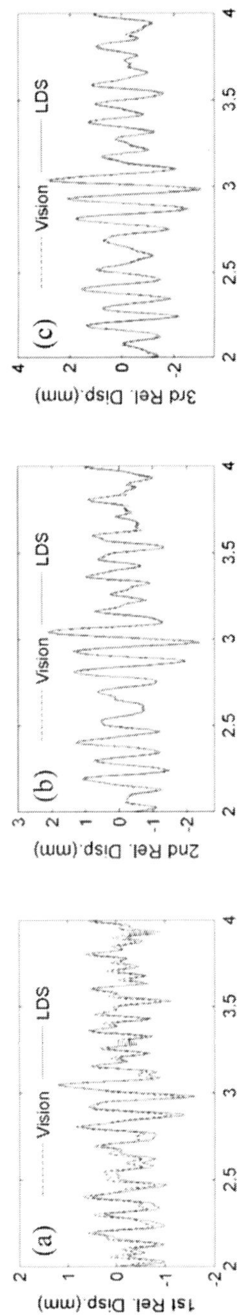

Figure 8. Comparisons of displacement relative to base floor between Vision (artificial target) and LDS: (**a**) 1st floor; (**b**) 2nd floor; (**c**) 3rd floor.

It is noted that the scaling factor for the vision sensor in the testing is 1.338 mm/pixel, meaning that the expected maximum error is 0.669 mm from a pixel-level template matching. However, since an upsampling factor of κ=20 is selected to achieve a 0.05 subpixel accuracy, the vision sensor can still accurately capture the small relative displacements ranging from 0 to 3 mm.

Next, instead of tracking the artificial targets, the four bolt connections on the frame structure in Figure 5, are used as the tracking targets. Figure 9 shows the displacement comparison of each floor by Vision (natural target) and LDS, respectively.Figure 10 plots the 1st, 2nd, and 3rd floor displacements relative to the base floor. Again, satisfactory agreements between Vision (natural target) and LDS can be achieved.

To quantify the accuracy and precision of the vision sensor, error analysis is performed using the normalized root mean squared error (NRMSE)

$$\mathrm{NRMSE}=\frac{\sqrt{\frac{1}{n}\sum_{i=1}^{n}(x_i-y_i)^2}}{y_{\max}-y_{\min}}\times 100\% \tag{8}$$

where n = number of measurement data; xi and yi=ith displacement data at time ti, measured by the vision sensor and the LDS, respectively; and ymax=max(yi), ymin=min(yi).

Table 3 tabulates the NRMSE errors of the displacement measurements in Figure 7 and Figure 9. It is concluded that vision sensor demonstrates a high accuracy by tracking either artificial targets or natural targets, with a maximum NRMSE error of 0.6%. By tracking natural targets without requiring artificial targets installed on fixed locations on the structure, the vision sensor provides the flexibility to easily change locations for displacement measurement, thus further facilitating the testing process.

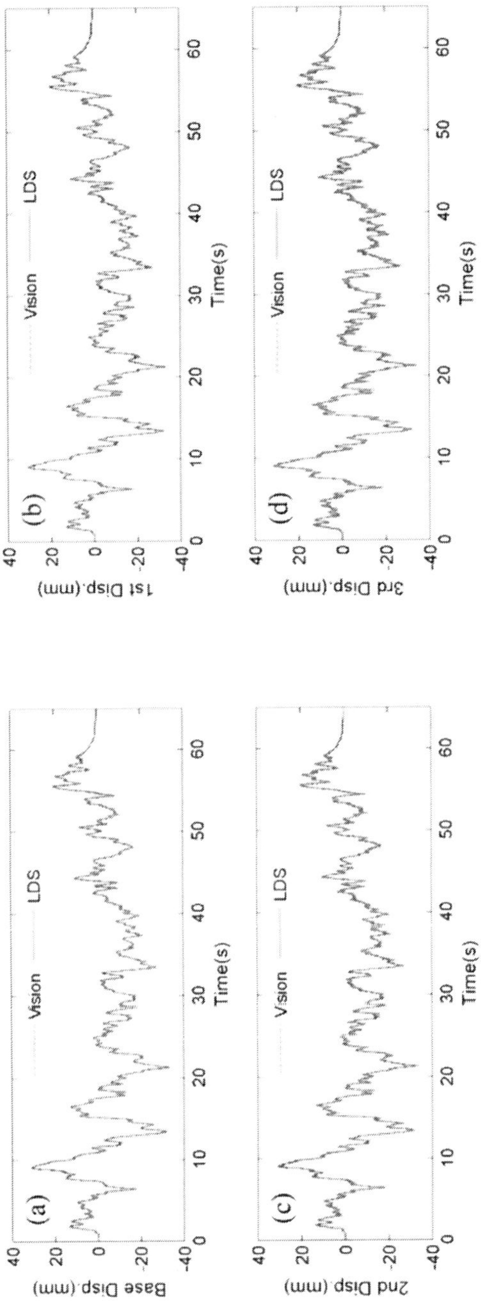

Figure 9. Displacement comparisons between Vision (natural target) and LDS: (**a**) Base floor; (**b**) 1st floor; (**c**) 2nd floor; (**d**) 3rd floor.

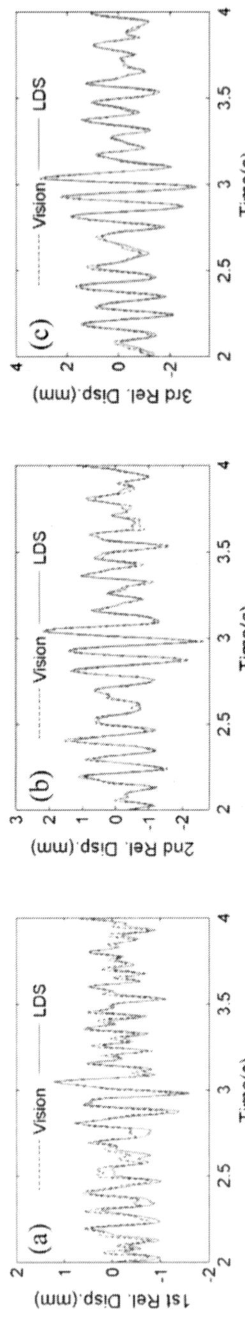

Figure 10. Comparisons of displacement relative to base floor between Vision (natural target) and LDS: (**a**) 1st floor; (**b**) 2nd floor; (**c**) 3rd floor.

Table 3. Measurement errors: normalized root mean squared error (NRMSE) (%).

Floor	Vision Sensor	
	Artificial Target	Natural Target
Base	0.39	0.60
1st	0.28	0.45
2nd	0.27	0.35
3rd	0.18	0.32

4. FIELD TESTS

To evaluate the performance of the vision sensor in realistic field environments, field tests are carried out on two bridges. Specifically, the time-domain performance is evaluated through field tests of a railway bridge by comparing with reference LVDT, and the frequency-domain performance is evaluated through field tests of a pedestrian bridge by comparing with the conventional accelerometer. A sampling rate of 150 frames per second with a resolution of 640 × 512 pixel was adopted.

4.1. Field Test of a Railway Bridge

In collaboration with the Transportation Technology Center, Inc. (TTCI), field measurements are carried out on a state-of-the-art hybrid composite bridge, which is one of the test-bed bridges in TTCI, Colorado. As shown in Figure 11a, the bridge is 12.8 m long. The train used for the testing has one locomotive and 15 freight cars. Figure 11b shows the artificial target and natural target on the bridge. The video camera is fixed on a tripod and set up at a remote location away from the bridge. This field tests focused on the measurement of the vertical displacement at the mid-span point by the vision sensor. As a reference sensor, a conventional contact-type displacement sensor, *i.e.*, a LVDT, is installed on the mid span of the bridge with one end connected to a stationary reference point on the ground through a string.

Figure 11. Field test of a railway bridge: (**a**) Displacement measurement under moving trainloads; (**b**) Artificial target and natural target.

Often in the field, it is difficult to find a location close to the structure to set up the vision sensor system, thus requiring to evaluate performance of vision sensor at different remote distances. Table 4 summarizes two of the field testing cases. Test T1 is conducted with a train speed of 40.23 km/h (or 25 mph) and with the vision sensor system placed 30.48 m (or 100 ft) away from the bridge, and Test T2 with a train speed of 64.36 km/h (or 40 mph) and with the sensor 60.96 m (or 200 ft) away from the bridge. It is noteworthy that the camera optical axis is tilted by small angles (2° for T1 and 1° for T2) with respect to the normal direction of the bridge surface, the errors from which are acceptable based on the scaling factor discussions inSection 2.1.

Table 4. Test cases.

Test	Measurement Distance (m)	Camera Tilt Angle (°)	Train Speed (km/h)	Scaling Factor (mm/pixel)
T1	30.48	2	40.23	1.90
T2	60.96	1	64.36	3.83

Figure 12 and Figure 13 plot respectively the displacement time histories for tests T1 and T2 by the three sensor systems, namely, the LVDT, Vision (artificial target), and Vision (natural target). In general, the measurements agree well with one another. Specifically, the test results show that the measurement error increases as the measurement distance increases, mainly caused the increased difficulty in tracking either the artificial target or the natural target as measurement distance increases.

During the field tests, two other problems are recognized, which would also contribute to the measurement errors. Firstly, the camera vibrations caused by moving-train-induced ground motion can affect the measurement accuracy when the camera is placed far away from the measurement target and zoom lenses magnifies not only the images but also the camera vibration. This problem becomes more serious for the proposed compact and portable vision-based displacement sensor system, because it is impossible to utilize stable concrete camera base fixture to avoid the micro camera vibration. The second problem is the heat haze that occurs when the air is heated, non-uniformly, by the high ambient temperature during the field testing. The non-uniformly heated air causes variation in its optical reflection index, resulting in image distortion, which would cause more measurement errors as the measurement distance increases, because the air thickness between the target object and the lens of the camera becomes large.

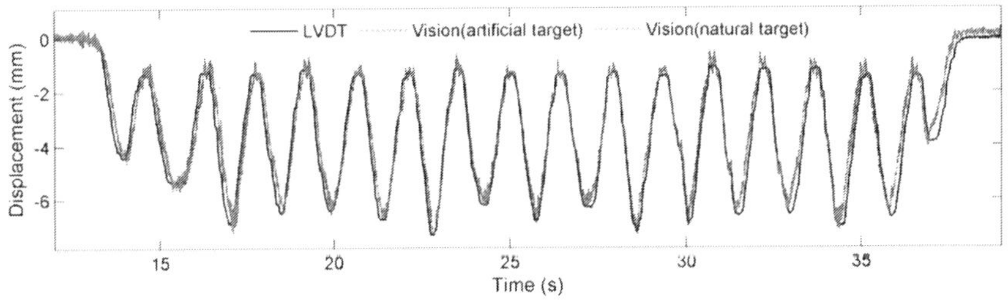

Figure 12. Comparison of displacements: Test T1.

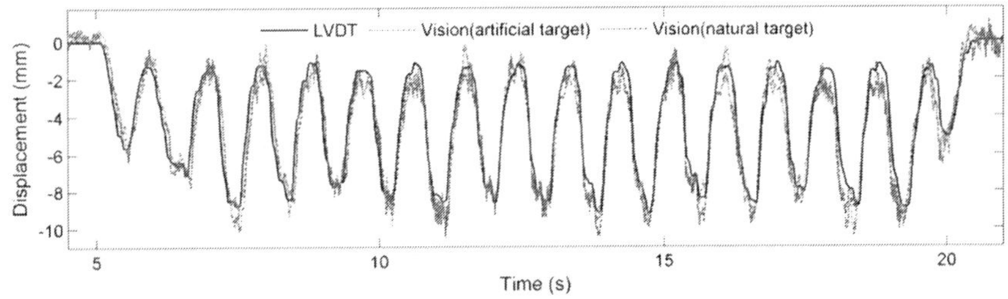

Figure 13. Comparison of displacements: Test T2.

4.2. Field Test of a Pedestrian Bridge

The Streicker Bridge is a pedestrian bridge located on the Princeton University campus, NJ, USA. The bridge has a main span and four approaching legs. The main span is a deck-stiffened arch and the legs are curved continuous girders supported by steel columns [33]. This field tests are to study the performance of the vision sensor in frequency domain. Two sets of dynamic loading tests are carried out on the third span of the southeast leg. As shown in Figure 14, one artificial target and one accelerometer (Model#W352C67 by PCB PIEZOTRONICS Inc.: Depew, NY, USA) are installed on the mid span. It is noted that the camera optical axis is tilted by an approximate angle of 15° with respect to the normal direction of the bridge surface, However, in this field test, due to the large height between ground and the bridge bottom surface, it is very difficult to install a reference LVDT to compare the accuracy of the measured displacement time histories by the vision sensor.

First, in order to apply dynamic loads with broadband frequency contents to the bridge, a group of pedestrians ran on the bridge deck randomly with different, varying speeds, rhythms and directions without any particular pattern. Figure 15shows the displacement measurement from the vision sensor together with the power spectral density (PSD) result. Figure 16 plots the acceleration measurement from the accelerometer and the corresponding PSD result. By comparing the results, one dominant frequency can be clearly identified as 3.08 Hz from both sensors, as well as two higher frequencies of 3.68 Hz and 4.47 Hz, respectively.

Figure 14. Field test: (**a**) Streicker Bridge; (**b**) Artificial target.

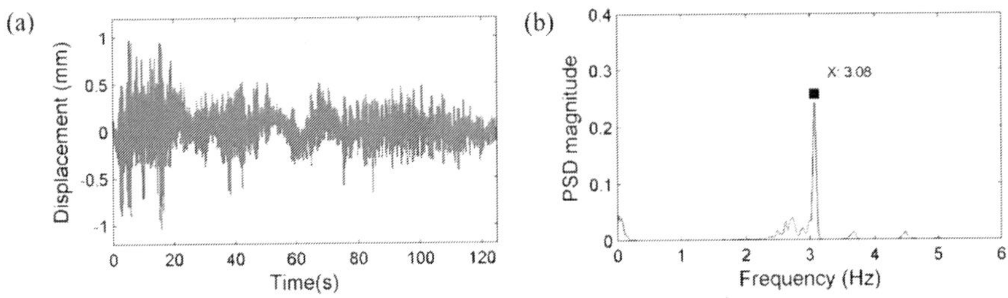

Figure 15. Randomly running of pedestrians: (**a**) Displacement by the vision sensor; (**b**) corresponding PSD.

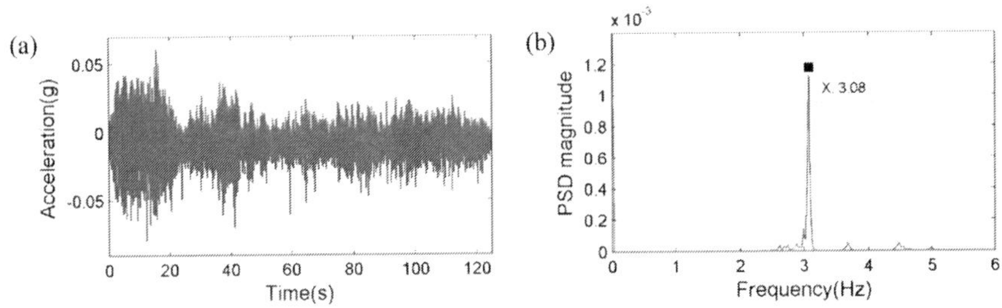

Figure 16. Randomly running of pedestrians: (**a**) Acceleration measurement; (**b**) Corresponding PSD.

Secondly, the pedestrian participants jumped on the mid span of the bridge deck synchronically with a frequency of around 3 Hz, which is close to the estimated first natural frequency of the bridge. Figure 17 and Figure 18 plot the displacement and acceleration time histories obtained respectively from the vision sensor and the accelerometer, together with corresponding PSD results. Again, the identified frequencies based on the two sensors show excellent agreement. Therefore, it is concluded that the same frequency components can be accurately obtained from the vision sensor.

5. CONCLUSIONS AND FUTURE WORK

In this study, a vision sensor system is developed for remote measurement of structural displacements based on an advanced subpixel template matching technique, namely, the upsampled cross correlation by means of Fourier transform. Comprehensive experiments, including a shaking table test and two bridge field tests, are carried out to investigate its performance. The following conclusions can be drawn:

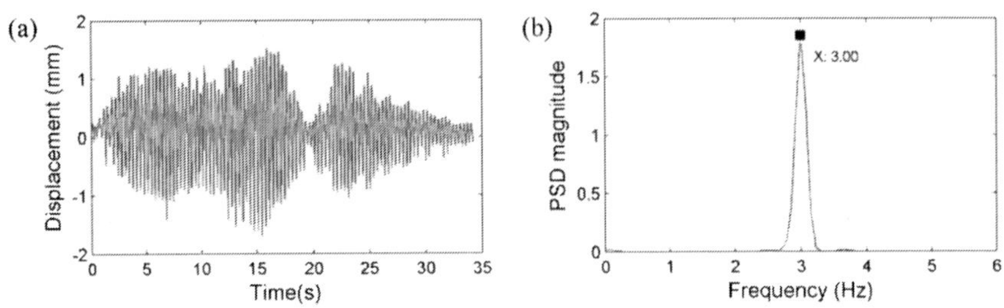

Figure 17. Jumping of pedestrians: (**a**) Displacement by the vision sensor; (**b**) Corresponding PSD.

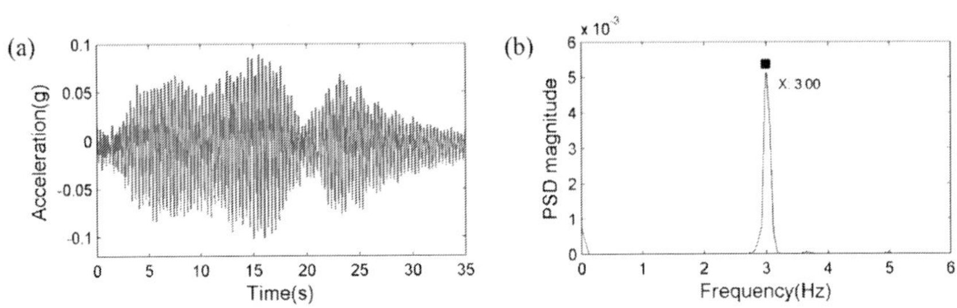

Figure 18. Jumping of pedestrians: (**a**) Acceleration measurement; (**b**) Corresponding PSD.

1) As a significant advantage of the proposed vision sensor, better subpixel resolution can be easily achieved by adjusting the upsampling factor. Thus structural vibrations smaller than 1 mm can be accurately measured.

2) From the shaking table test of a frame structure, satisfactory agreements are observed between the multi-point displacement time histories measured at all floors by one camera by tracking bolt

connections on the structure surface and those by four laser displacement sensors.

3) In realistic field environments, the time-domain performance of the vision sensor is further confirmed through field tests of a railway bridge during train passing; and the frequency-domain performance is validated through field tests of a pedestrian bridge subjected to dynamic loading.

By tracking existing natural targets on the structure surface, the vision sensor developed in this study provides the flexibility to easily change locations for displacement measurement. The availability of such as a remote sensor will facilitate cost-effective monitoring of civil engineering structures.

As part of our plan to improve the measurement accuracy in uncontrolled outdoor field environments, we are working to reduce the errors caused by the heat haze and camera vibrations. Moreover, in order to study the potentials of the vision sensor for structural health monitoring, we are building simply beam specimens with different kinds of damages. The aim is to evaluate the effectiveness of the vision sensor with respect to: (1) measuring full-field displacement responses using one camera; and (2) extracting modal information (natural frequencies and mode shapes) from measured multi-point displacements and detecting damages in beam structures.

ACKNOWLEDGMENTS

The field test at the railway bridge is partially supported by TTCI through Federal Railroad Administration. Special thanks go to Richard Joy, Dingqing Li, and Duane Otter at TTCI for their help in planning and executing the field test. The field test of the pedestrian bridge is organized by B. Glisic at Princeton University. The authors are thankful for their contributions.

AUTHOR CONTRIBUTIONS

The vision sensor system is developed by D.M. Feng, under the supervision of M. Feng. Experiment planning, setup and measurement of laboratory and field tests are conducted by D.M. Feng, M. Feng, E. Ozer and Y. Fukuda. The article is written by D.M. Feng and M. Feng.

CONFLICTS OF INTEREST

The authors declare no conflict of interest.

REFERENCES

1. Fukuda, Y.; Feng, M.Q.; Shinozuka, M. Cost-Effective vision-based system for monitoring dynamic response of civil engineering structures. *Struct. Control Health Monit.* **2010**, *17*, 918–936.

2. Fukuda, Y.; Feng, M.Q.; Narita, Y.; Kaneko, S.; Tanaka, T. Vision-Based displacement sensor for monitoring dynamic response using robust object search algorithm. *IEEE Sens. J.* **2013**, *13*, 4725–4732.

3. Ribeiro, D.; Calçada, R.; Ferreira, J.; Martins, T. Non-Contact measurement of the dynamic displacement of railway bridges using an advanced video-based system. *Eng. Struct.* **2014**, *75*, 164–180.

4. Kohut, P.; Holak, K.; Uhl, T.; Ortyl, Ł.; Owerko, T.; Kuras, P.; Kocierz, R. Monitoring of a civil structure's state based on noncontact measurements. *Struct. Health Monit.* **2013**, *12*, 411–429.

5. Gentile, C.; Bernardini, G. An interferometric radar for non-contact measurement of deflections on civil engineering structures: Laboratory and full-scale tests. *Struct. Infrastruct. Eng.* **2009**, *6*, 521–534.

6. Nassif, H.H.; Gindy, M.; Davis, J. Comparison of laser doppler vibrometer with contact sensors for monitoring bridge deflection and vibration. *NDT E Int.* **2005**, *38*, 213–218.

7. Casciati, F.; Fuggini, C. Monitoring a steel building using gps sensors. *Smart Struct. Syst.* **2011**, *7*, 349–363.

8. Casciati, F.; Wu, L. Local positioning accuracy of laser sensors for structural health monitoring. *Struct. Control. Health Monit.* **2013**, *20*, 728–739.

9. Kim, S.-W.; Jeon, B.-G.; Kim, N.-S.; Park, J.-C. Vision-Based monitoring system for evaluating cable tensile forces on a cable-stayed bridge. *Struct. Health Monit.* **2013**, *12*, 440–456.

10. Schumacher, T.; Shariati, A. Monitoring of structures and mechanical systems using virtual visual sensors for video analysis: Fundamental concept and proof of feasibility. *Sensors* **2013**, *13*, 16551–16564.

11. Song, Y.-Z.; Bowen, C.R.; Kim, A.H.; Nassehi, A.; Padget, J.; Gathercole, N. Virtual visual sensors and their application in structural health monitoring. *Struct. Health Monit.* **2014**, *13*, 251–264.

12. Busca, G.; Cigada, A.; Mazzoleni, P.; Zappa, E. Vibration monitoring of multiple bridge points by means of a unique vision-based measuring system. *Exp. Mech.* **2014**, *54*, 255–271.

13. Santos, C.A.; Costa, C.O.; Batista, J.P. Calibration methodology of a vision system for measuring the displacements of long-deck suspension bridges. *Struct. Control. Health Monit.* **2012**, *19*, 385–404.

14. Debella-Gilo, M.; Kääb, A. Sub-Pixel precision image matching for measuring surface displacements on mass movements using normalized cross-correlation. *Remote Sens. Environ.* **2011**, *115*, 130–142.

15. Lee, J.J.; Shinozuka, M. A vision-based system for remote sensing of bridge displacement. *NDT&E Int.* **2006**, *39*, 425–431.

16. Lee, J.-J.; Ho, H.-N.; Lee, J.-H. A vision-based dynamic rotational angle measurement system for large civil structures.*Sensors* **2012**, *12*, 7326–7336.

17. Park, H.; Kim, J.; Kim, J.; Choi, S.; Kim, Y. A new position measurement system using a motion-capture camera for wind tunnel tests. *Sensors* **2013**, *13*, 12329–12344.

18. Wu, L.-J.; Casciati, F.; Casciati, S. Dynamic testing of a laboratory model via vision-based sensing. *Eng. Struct.* **2014**, *60*, 113–125.

19. Sładek, J.; Ostrowska, K.; Kohut, P.; Holak, K.; Gąska, A.; Uhl, T. Development of a vision based deflection measurement system and its accuracy assessment. *Measurement* **2013**, *46*, 1237–1249.

20. Feng, M.; Fukuda, Y.; Feng, D.; Mizuta, M. Nontarget vision sensor for remote measurement of bridge dynamic response. *J. Bridg. Eng.* **2015**.

21. Pan, B.; Xie, H.-M.; Xu, B.Q.; Dai, F.L. Performance of sub-pixel registration algorithms in digital image correlation.*Meas. Sci. Technol.* **2006**, *17*, 1615.

22. Foroosh, H.; Zerubia, J.B.; Berthod, M. Extension of phase correlation to subpixel registration. *IEEE Trans. Image Process.* **2002**, *11*, 188–200.

23. Berenstein, C.A.; Kanal, L.N.; Lavine, D.; Olson, E.C. A geometric approach to subpixel registration accuracy. *Comput. Vis. Graph. Image Process.* **1987**, *40*, 334–360.

24. Pilch, A.; Mahajan, A.; Chu, T. Measurement of whole-field surface displacements and strain using a genetic algorithm based intelligent image correlation method. *J. Dyn. Syst. Meas. Control* **2004**, *126*, 479–488.

25. Li, L.; Chen, Y.; Yu, X.; Liu, R.; Huang, C. Sub-pixel flood inundation mapping from multispectral remotely sensed images based on discrete particle swarm optimization. *ISPRS J. Photogramm. Remote Sens.* **2015**, *101*, 10–21.

26. Bruck, H.A.; McNeill, S.R.; Sutton, M.A.; Peters, W.H., III. Digital image correlation using newton-raphson method of partial differential correction. *Exp. Mech.* **1989**, *29*, 261–267.

27. Davis, C.Q.; Freeman, D.M. Statistics of subpixel registration algorithms based on spatiotemporal gradients or block matching. *Opt. Eng.* **1998**, *37*, 1290–1298.

28. Hijazi, A.; Friedl, A.; Kähler, C.J. Influence of camera's optical axis non-perpendicularity on measurement accuracy of two-dimensional digital image correlation. *Jordan J. Mech. Ind. Eng.* **2011**, *5*, 373–382.

29. Zhang, Z. A flexible new technique for camera calibration. *IEEE Trans. Pattern Anal. Mach. Intell.* **2000**, *22*, 1330–1334.

30. Dworakowski, Z.; Kohut, P.; Gallina, A.; Holak, K.; Uhl, T. Vision-Based algorithms for damage detection and localization in structural health monitoring. *Struct. Control Health Monit.* **2015**.

31. Guizar-Sicairos, M.; Thurman, S.T.; Fienup, J.R. Efficient subpixel image registration algorithms. *Opt. Lett.* **2008**, *33*, 156–158.

32. Tian, Q.; Huhns, M.N. Algorithms for subpixel registration. *Comput. Vis. Graph. Image Process.* **1986**, *35*, 220–233.

33. Abdel-Jaber, H.; Glisic, B. Analysis of the status of pre-release cracks in prestressed concrete structures using long-gauge sensors. *Smart Mater. Struct.* **2015**, *24*.

Chapter 4

Research on Workpiece Sorting System Based on Machine Vision Mechanism

Juan Yan, Huibin Yang

College of Mechanical Engineering, Shanghai University of Engineering Science, Shanghai, China

ABSTRACT

This paper describes industrial sorting system, which is based on robot vision technology, intro- duces main image processing methodology used during development, and simulates algorithm with Matlab. Besides, we set up image processing algorithm library via C# program and realize recognition and location for regular geometry workpiece. Furthermore, we analyze camera model in vision algorithm library, calibrate the camera, process the image series, and resolve the identify problem for regular geometry workpiece with different colours.

KEYWORDS

Machine Vision, Industrial Robot, Target Recognition, Image Processing

1. INTRODUCTION

With the development of society, enterprises have become increasingly demanding on production automation, so that more and more industrial robots have been used in automated production lines. For example, in order to enhance operational efficiency in modern logistics distribution centers and to improve the efficiency of large- scale centralized distribution to reduce operating costs, as the core equipment distribution center and the main operating procedures, sorting systems and sorting operation efficiency and technology are increasingly by theorists and engineering concern. Another example is the tool change robot in the machining center. Compared to ma- nual tool change method, this tool works more quickly and more accurately, greatly reduces tool change time and increases production efficiency. In the automotive industry, due to the auto parts more complex vehicle bodies, in many cases on the body cannot use manual welding. Currently, automobile manufacturers utilize a lot of body welding robots to improve product quality and to reduce the labor intensity. Visibly, doing the work on the production line with industrial robots is an inevitable trend.

Machine vision technology uses cameras and computers to simulate human vision features, which are widely used in electronics, aerospace, automotive and automotive parts manufacturing industry, the pharmaceutical industry and electronics fields [1] . In the traditional production line, sorting artifacts with industrial robots usually uses teaching methods or offline programming operations. All the actions and placements, etc. should be strictly preset. Once the working environment or condition changes, it will affect the efficiency and accuracy of crawling robot, unable to meet the high-speed production of large quantities. Introduction of machine vision in industrial production can greatly improve productivity and reduce costs. Industrial machine vision technology based on sorting system with high detection speed, high reliability, and high real-time, compared to traditional mechanical sorting, is more intelligent, more efficient and has other irreplaceable advantages.

This paper researches the sorting system based on the machine vision and the methods of camera calibration. The straight linear method is used and the system is simulated by Matlab. It also studied the correlation of image processing algorithms, combined thresholding proposed centroid location method to solve the different colors of regular geometric shape of the workpiece recognition problem. Threshold segmentation and edge detection method can lock the workpiece area; centroid location can accurately locate the geometric center of the workpiece. And this integrated approach can effectively solve the regular geometric shape of the workpiece sorting problem.

2. SORTING SYSTEM BASED ON MACHINE VISION HARDWARE ARCHITECTURE ARTIFACTS

In this paper, the mechanical axis motion platform, is built on the work of sorting machine vision systems. Overall system structure is shown in Figure 1.

The hardware platform consists of mechanical three-axis motion platform, camera platforms, PC and motion control unit, four major components. Hardware platform package sorting system is shown inFigure 2.

Mechanical three-axis motion platform: Using screw-nut mechanism, while the linear guide linear motion needs as a guide and support. The system uses a servo motor drive: although the control unit is complex, the high repeat accuracy, no step phenomenon and no cumulative error [2] .

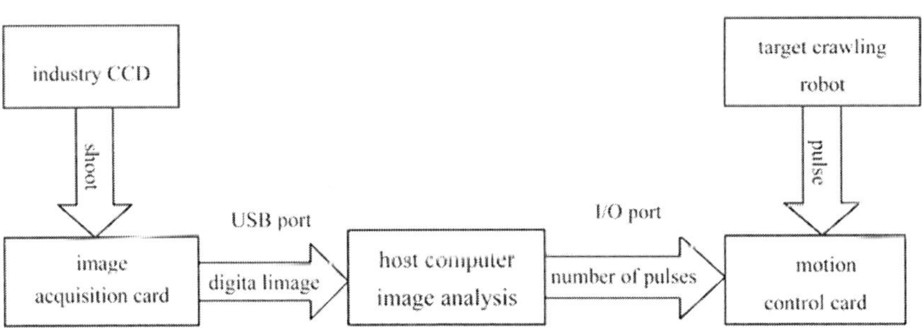

Figure 1. Piece sorting system structure based on machine vision.

Camera Platform: Camera platform unit is mainly constructed by MV-2000UC camera and light source com- ponents. Hanging bracket role monocular camera is to obtain the experimental stage of the video image of the workpiece. White LED surface light source provides light for digital cameras captured images. The light will be fixed at the top of the workpiece in order to eliminate the shadow of the workpiece itself.

PC: Using existing laboratories as a PC, since the development environment is good, fast, easy to operate and low cost. At that time with respect to the IPC, its anti-interference ability and stability is poor, such as teaching or laboratory for stable work environment occasions. PC camera platform receives the experimental stage to obtain a video image of the workpiece using the vision system to identify the target species, the centroid of the workpiece is then calculated, and finally the image according to the relationship between the coordinate system and the object coordinate system, the calculated position of the target relative and direction, and then passes the information parameter to the control unit.

Figure 2. Workpiece sorting system based on machine vision hard- ware platform.

Motion Control unit: mainly control the composition of the cabinet and robots. It is responsible for the para- meters analyzing, then the robot-related operations and finally the completion of the target workpiece sorting grab and place.

3. CAMERA CALIBRATION

Camera calibration is one of the important parts in machine vision [3] . Calibration includes calibration camera and visual systems. First of all based model of the camera, and then to establish the relationship between the coordinate system, and finally through the calibration can establish the correspondence between image pixels and the spatial three-dimensional coordinates [4] . Workpiece sorting system is based on machine vision. The camera calibration as a key link to other related work is a prerequisite for the smooth conduct of the direct impact on subsequent sorting processes. Camera calibration process is based on the model of the camera imag- ing process to solve internal and external camera parameters, and these parameters determine the correspon- dence between the image pixel point and three-dimensional coordinate space.

There are many ways in camera calibration, the current methods commonly used in machine vision are direct linear method, perspective projection matrix method, Zhang Zhengyou France, the two-step camera calibration method. For the facts of relatively stable laboratory and low interference environment, this paper chooses the direct linear method, which is characterized by convenient computing but less accuracy. It is difficult to be pre- cise shooting environment complex self-correcting. DLT method called direct linear method, by Abdel-Aziz and Karara and was first proposed in 1971 [5] . This method does not consider the non-linear distortion of the camera, getting the relevant parameters of the camera by directly solving a set of linear equations. It acquires pinhole imaging model for the study, ignoring specific intermediate imaging process and using a 3 × 4 order matrix to represent a direct correspondence between the point and the two-dimensional space object image point. The linear transformation matrix is with only a difference of a perspective matrix scale factor. Providing space point $(X,Y,Z)\ P$ coordinates in the camera coordinate system of (x,y,z), the camera will capture the image on the CCD imaging plane, which is provided for the image point coordinates (x,y) corresponding to the image pixel coordinates (u,v).

By the world coordinate system to the camera coordinate system, using the following formula rotation—trans- lation conversion formula described:

$$\begin{bmatrix} x \\ y \\ z \\ 1 \end{bmatrix} = \begin{bmatrix} R & T \\ 0 & 1 \end{bmatrix} \begin{bmatrix} X \\ Y \\ Z \\ 1 \end{bmatrix} \tag{1}$$

where in R is a rotation matrix, T is the translation matrix. In the case where distortion is not considered (i.e., an ideal perspective projection), the image coordinates of the camera coordinate system is described by the following formula:

$$s\begin{bmatrix} X \\ Y \\ 1 \end{bmatrix} = \begin{bmatrix} f & 0 & 0 \\ 0 & f & 0 \\ 0 & 0 & 1 \end{bmatrix} \begin{bmatrix} x \\ y \\ z \end{bmatrix} \tag{2}$$

In the image processing, is often used in image coordinate of the pixel as a unit, so to convert the image coordinate system to the coordinate system of the pixel, i.e.

$$\begin{bmatrix} u \\ v \\ 1 \end{bmatrix} = \begin{bmatrix} \frac{1}{d_X} & 0 & 0 \\ 0 & \frac{1}{d_Y} & 0 \\ 0 & 0 & 1 \end{bmatrix} \begin{bmatrix} X \\ Y \\ 1 \end{bmatrix} \tag{3}$$

Without considering the various types of the imaging by the specific process, can be obtained:

$$s\begin{bmatrix} u \\ v \\ 1 \end{bmatrix} = \begin{bmatrix} r_{11} & r_{12} & r_{13} & r_{14} \\ r_{21} & r_{22} & r_{23} & r_{24} \\ r_{31} & r_{32} & r_{33} & r_{34} \end{bmatrix} \begin{bmatrix} X \\ Y \\ Z \\ 1 \end{bmatrix} \tag{4}$$

where, (X,Y,Z) of the three-dimensional world coordinate space points, (u,v) coordinates of the corres- ponding pixel, r_{ij} is a perspective view of elements of the transformation matrix. It contains three equations, finishing eliminates, obtain the following two linear equations on people r_{ij}:

$$r_{11}X + r_{12}Y + r_{13}Z + r_{34} - uXr_{31} - uYr_{32} - uZr_{33} = ur_{34} \quad (5)$$

$$r_{21}X + r_{22}Y + r_{23}Z + r_{24} - vXr_{31} - vYr_{32} - uZr_{33} = ur_{34} \quad (6)$$

This equation describes the relationship between the two three-dimensional world and the corresponding points between image points. If the three-dimensional world coordinates and the corresponding image coor- dinates are known, the transformation matrix regarded as unknown, then there will be a total of 12 unknowns. For two equations above each object point has, in general, can be set $r_{34} = 1$, then there will be a total of 11 unknowns. Taking six goals can get 12 points a overdetermined equation Using the least squares method des- cribed above can easily find solution of linear equations [6] . Using more points, we make much more than the number of equations number of unknowns, using the least squares method for solving can reduce the impact of errors caused by the perspective transformation matrix obtained after its decomposition can get cameras inside and outside the parameters.

Based on the direct linear method was calibrated to the camera. Put the calibration paper on a suitable position in the robot coordinate system, as shown in Figure 3. Cross figure crosshairs for the world coordinate system. Only a simple translation of the relationship is needed between it and the robot base coordinate system, i.e., calculating the target position in the world coordinate system, then subtracting a certain shift amount and finally obtaining the target based on the robot base in the standard position.

Corner extraction algorithm using corner point data is shown in Table 1.

On the Matlab model with a blue asterisk plotted coordinates of each point (real value), the table on each corner point coordinates a left multiplication result (calculated) M-1 obtained plotted with a green plus sign, as shown in Figure 4.

Figure 3. Camera calibration chart based DLT.

The verification result from Figure 4 is that, by using direct linear method in this system, the calibration results is accurate, easy to calculate and close to actual value. Besides, the calibration results are also good, which provides a reliable foundation for accurate positioning of the workpiece in a sorting system.

4. THRESHOLDING SEGMENTATION

This design uses machine vision technology, the different colors of the workpiece intelligent sorting rules. When the camera calibration is completed, after the pixel coordinates to the scenecoordinate correspondence to obtain the target, will be on the video image is processed by computer.

First, the camera image acquired in the target workpiece is detected. Target detection is extracted from the image portion of interest, i.e., the image segmentation; the image processing to the image analysis is a key step. At present, a lot of image segmentation methods, such as thresholding method, edge detection and region extraction method. In this paper, we use the thresholding method and edge detection method to detect the target workpiece.

Thresholding as the most common color distinction is a region-based image segmentation technique. Briefly, image segmentation threshold is transformed to a color image to grayscale, the first grayscale range of an image to determine the gray value threshold, then the image of the gray value of each pixel with this threshold value, and are classified according to the size of the result of the comparison, based on the threshold values of all the pixels in the image is divided into two areas.

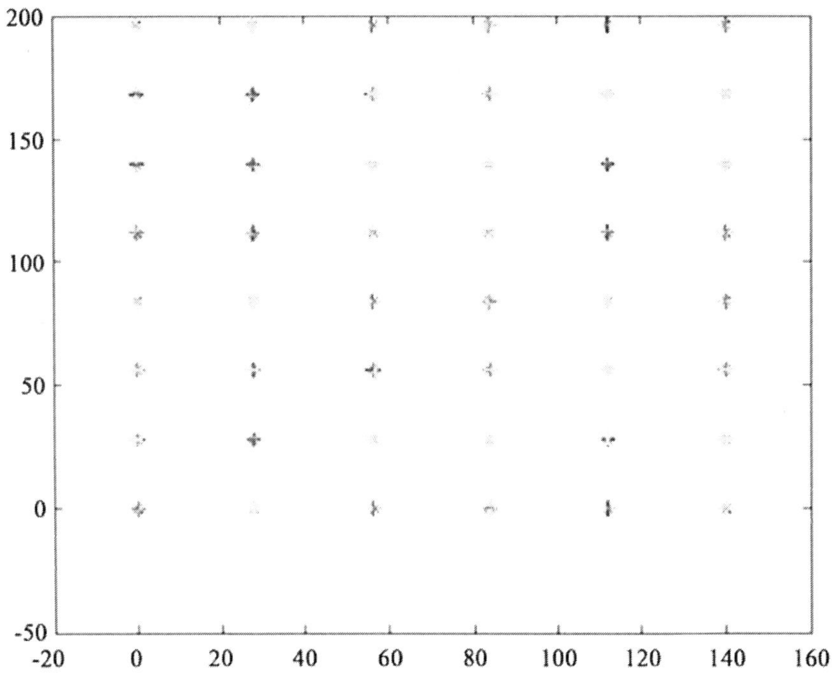

Figure 4. Camera calibration results based DLT.

Table 1. Angular point data table.

Angular point data table				
(u, v)	Column 1	Column 2	Column 3	Column 4
Line 1	(234.5, 168)	(261.5, 166.5)	(289.5, 166.5)	(316.5, 166.5)
Line 2	(235.5, 194.5)	(262.5, 194.5)	(289.5, 193.6)	(316.5, 193.5)
Line 3	(235.5, 221.5)	(262.5, 221.5)	(289.5, 221.5)	(316.5, 220.5)
Line 4	(235.5, 248.5)	(262.5, 248.5)	(290, 248.5)	(316.5, 247.5)
Line 5	(236.5, 275.5)	(262.5, 275.5)	(290.5, 275.5)	(317, 275)
Line 6	(236.5, 302.5)	(263.5, 302.5)	(290.5, 302.5)	(317.5, 302.5)
(u, v)	Column 5	Column 6	Column 7	Column 8
Line 1	(344, 166)	(371, 166)	(397.5, 165.5)	(425, 165.5)
Line 2	(343, 193)	(370, 194)	(397.5, 192.5)	(425.5, 192.5)
Line 3	(343.5, 220.5)	(370.5, 220.5)	(398.5, 220.5)	(425.5, 219.5)
Line 4	(344, 247.5)	(371, 247.5)	(398.5, 247.5)	(425.5, 247.5)
Line 5	(344.5, 247.5)	(371.5, 274.5)	(398.5, 274.5)	(425.5, 274.5)
Line 6	(344, 301.5)	(371.5, 301.5)	(398.5, 301.5)	(398.5, 301.5)

A pair of the original image $f(x,y)$ is taken to define a single threshold value T of the divided

$$\text{image is}: g(x,y)=\begin{cases}1 & f(x,y)>T\\0 & f(x,y)\leq T\end{cases} \tag{7}$$

In the actual case, the difference between the image of the target workpiece and background brightness is small, so the threshold needs to be adjusted manually. Figure 5 shows the use of maximum between-class variance method combined with manual adjustment method in Matlab to finalize the process threshold.

From the segmentation results can be seen, when the threshold of T is 100, it's the best segmentation, can accurate segmentation of the target range of workpiece, which proves Ostu method combined with manual adjustment is the most suitable way for this system.

5. DETERMINE THE CENTROID

In the sorting operation of the workpiece, the workpiece needs to be identified for positioning, the general centroid coordinates describing the location information of the workpiece. But the calculation method for multi- target centroid such as starting algorithms geometric center of the target group, the calculation is more com- plicated; through rows of pixels accumulating and averaging roughly centroid location, can not meet the precise positioning. According to the characteristics identified herein workpiece shape rules proposed rule graphics centroid location method [7].

When the function is applied to shape analysis, suppose there are binary function:

$$m_{pq}=\int_{-\infty}^{+\infty}\int_{-\infty}^{+\infty}x^{p}x^{q}f(x,y)\mathrm{d}x\mathrm{d}y \quad p,\, q\in N_0=\{0,1,\cdots\} \tag{9}$$

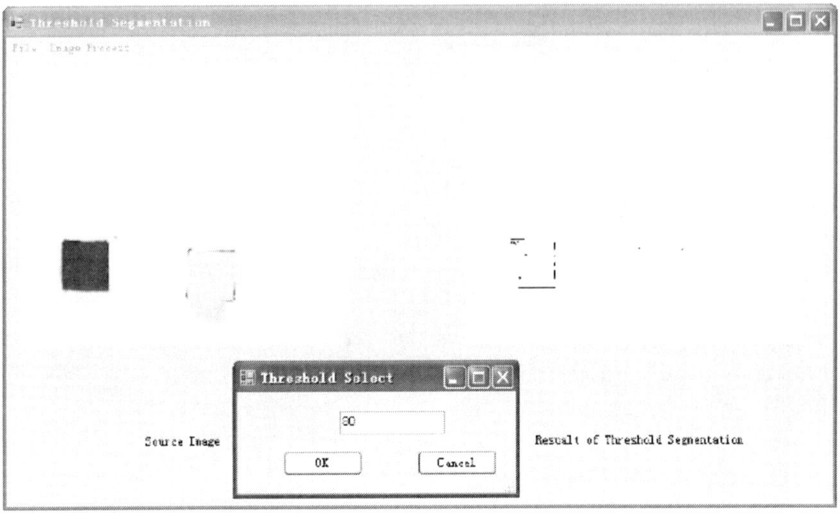

(a)

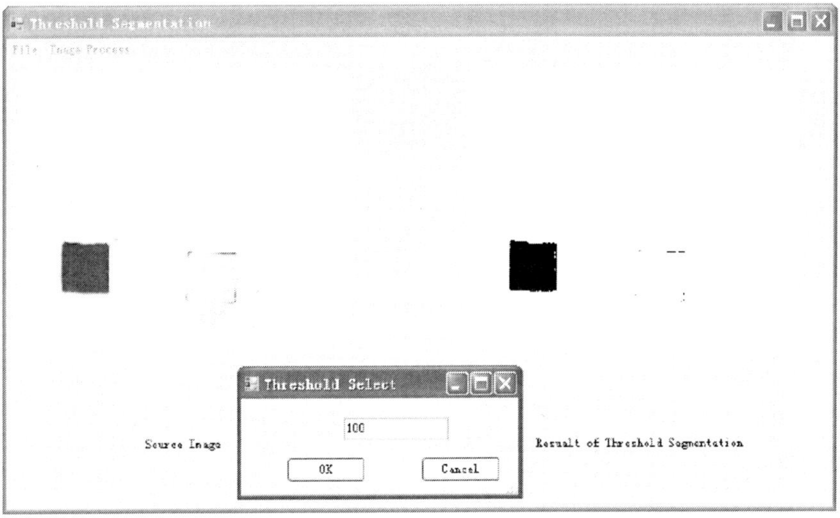

(b)

Figure 5. Threshold segmentation results. (a) When the threshold T to 80 threshold segmentation results; (b) When the threshold T to 100 threshold segmentation results.

Take $(p+q)$ as the moment of order. The moment for $M \times N$ digital image $f(i,j)$, $(p+q)$ is defined as:

$$m_{pq} = \sum_{i=0}^{M-1} \sum_{j=0}^{N-1} i^p i^q f(i,j) \tag{10}$$

0-order moment is the area of the sum of the image gray, m_{00} values represent object binary image, if used to standardize an order moments m_{10} and m_{01}, we get the coordinates of the object's center of gravity (i,j).

$$\begin{cases} \bar{i} = m_{10}/m_{00} = \sum_{i=0}^{M-1}\sum_{j=0}^{N-1} if(i,j) \Big/ \sum_{i=0}^{M-1}\sum_{j=0}^{N-1} f(i,j) \\ \bar{j} = m_{01}/m_{00} = \sum_{i=0}^{M-1}\sum_{j=0}^{N-1} jf(i,j) \Big/ \sum_{i=0}^{M-1}\sum_{j=0}^{N-1} f(i,j) \end{cases} \tag{11}$$

For binary images, grayscale only 0 and 1, the focus is the centroid. Finally, the first moment m_{10} and m_{01} is divided by zero-order moments to get the coordinates of the center of mass of the object:

$$(\bar{i}, \bar{j}) = (M_{10}/M_{00}, M_{01}/M_{00}) \tag{12}$$

After tests it is proved that the algorithm is simple and effective and is suitable for any graphics.

In this paper, the workpiece has been recognized as a rectangular, triangular, circular or a regular shape of the workpiece. The program is written in C#, in an ideal environment, workpiece recognition results are shown in Figure 6.

The rule workpiece centroid recognition results of Figure 6 demonstrates that, it is simple and efficient to adopt rule centroid locolization algorithm and is suitable for any graphic. Besides, programming is simple, recognition is speedy and recognition results are accurate.

6. CONCLUSIONS

This paper describes the main algorithm in the industrial sorting machine vision technology used in the system, using thresholding to detect the target work area and describe the target workpiece position information com- bined centroid location method. The recognition algorithm is programmed in MATLAB environment to conduct a simulation algorithm and it is developed using C# rule artifacts recognition software. According to the results obtained, to identify the robot reaches the top of the target number of pulses required to send to the motion control card, by motion control card for controlling the motor and the drive to achieve sorting purposes.

The results show that the practical application of technical methods discussed in this article is simple and effective. The sorting system identifies the type of workpiece correct rate of 100%, error of the target location is less than 5 mm, and further validates the feasibility of the system for robot.

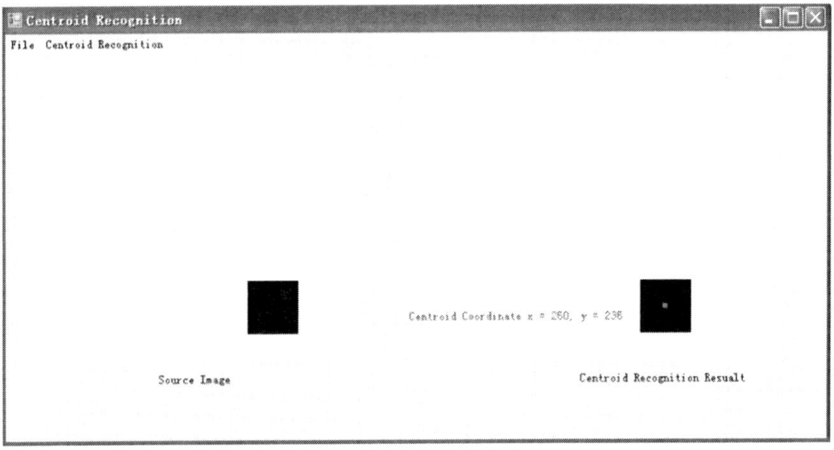

(a)

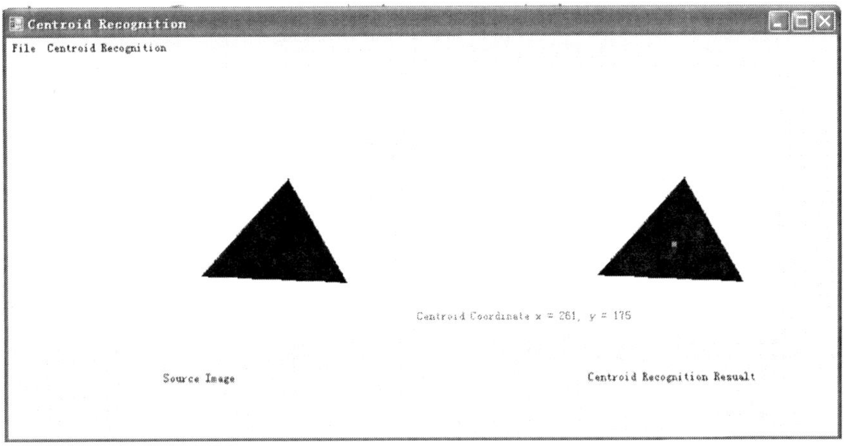

(b)

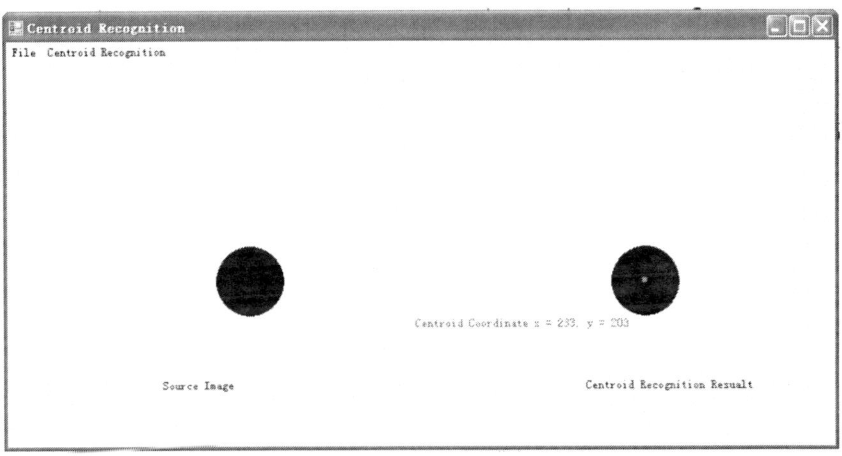

(c)

Figure 6. Rule workpiece centroid recognition results. (a) Rectangle centroid recognition results; (c) Triangle centroid recognition results; (c) Circular centroid recognition results.

REFERENCES

1. Zhuang, K.L., Wang, J.Z. and Zhou, J. (2011) Application of Machine Vision Technology in Terms of Angle Detection. Equipment Manufacturing Technology, 4, 9-11.

2. El Masry, G., Cubero, S. and Molto, E. (2012) In-Line Sourting of Irregular Potatoes by Using Automated Computer- Based Machine Vision System. Journal of Food Engineering, 112, 60-68.

3. Liu, Z.Y., Zhao, B. and Zou, F.S. (2012) Machine Vision Technology in the Sorting of the Workpiece. Computer Appli- cations and Software, 29, 87-91.

4. Quan, H.H. and Zhang, L.P. (2007) Research on Machine Vision-Based Assembly Line Parts Recognition. Modular Machine Tool & Automatic Manufacturing Technique, 12, 58-60.

5. Forsyth, D.A. and Ponce, J. (2002) Computer Vision: A Modern Approach. Prentice Hall, New Jersey.

6. Shapiro, R. (1976) Direct Linear Transformation Method for Three-Dimensional Cinematography. Research Quar- terly, 49, 197-205.

7. Ou, X.Y., Hou, X.Z. and Zheng, K. (2012) Machine Vision Based Automatic Teaching System of Six DOF Robot. Ma- nufacturing Automation, 34, 1-4.

Chapter 5

Vision-Based Steering Control, Speed Assistance and Localization for Inner-City Vehicles

Miguel Angel Olivares-Mendez [1,2,*], Jose Luis Sanchez-Lopez [1], Felipe Jimenez [3], Pascual Campoy [1], Seyed Amin Sajadi-Alamdari [2] and Holger Voos [2]

[1] Centre for Automation and Robotics (CAR), Universidad Politécnica de Madrid (UPM-CSIC), Calle de José Gutiérrez Abascal 2, 28006 Madrid, Spain

[2] Interdisciplinary Centre for Security, Reliability and Trust, SnT - University of Luxembourg, Weicker 1711, Luxembourg

[3] University Institute for Automobile Research (INSIA), Technical University of Madrid, INSIA, Campus Sur UPM, Carretera de Valencia km 7, 28031 Madrid, Spain

ABSTRACT

Autonomous route following with road vehicles has gained popularity in the last few decades. In order to provide highly automated driver assistance systems, different types and combinations of sensors have been presented in the literature. However, most of these approaches apply quite sophisticated and expensive sensors, and hence, the development of a cost-efficient solution still remains a challenging problem. This work proposes the use of a single monocular camera sensor for an automatic steering control, speed assistance for the driver and localization of the vehicle on a road. Herein, we assume that the vehicle is mainly traveling along a predefined path, such as in public transport. A computer vision approach is presented to detect a line painted on the road, which defines the path to follow. Visual markers with a special design painted on the road provide information to localize the vehicle and to assist in its speed control. Furthermore, a vision-based control system, which keeps the vehicle on the predefined path under inner-city speed constraints, is also presented. Real driving tests with a commercial car on a closed circuit finally prove the applicability of the derived approach. In these tests, the car reached a maximum speed of 48 km/h and successfully traveled a distance of 7 km without the intervention of a human driver and any interruption.

KEYWORDS

road vehicles; vision sensor; automatic steering control; speed assistance; computer vision; vision-based control; positioning; driver assistance vision-based system; partial automation

1. INTRODUCTION

Today the concept of autonomous driving is a very active field of research and first solutions of more and more advanced driver assistance systems are already available in recent models of commercial cars. As mentioned in [1], "the majority of the technologies required to create a fully autonomous vehicle already exist. The challenge is to combine existing automated functions with control, sensing and communications systems, to allow the vehicle to operate autonomously and safely". That report also presents a classification of the level of autonomy based on the capabilities provided by an autonomous system. The simplest system includes the human driver along with an electronic stability and cruise control, which is available in most of the new car models. The next level adds a driver assistance in which steering and/or acceleration is automated in specific situations like parking assistance and adaptive cruise control. The classification then continues with partial autonomy, in which the driver does not control the steering and/or acceleration, but can take over control again if it is required like in lane keeping. After that, there is the level of high autonomy in which the car system is able to operate autonomously in different sections of the journey and only gives the control back to the human driver in some specific dangerous situations. Finally, there is the level of full autonomy in which the vehicle is capable of driving an entire journey without human intervention. Herein, the vehicle must be able to provide all the following specific capabilities:

1) Self-localization in a map or on a predefined path.

2) Sensing the surrounding and identification of potential collisions.

3) Control the basic driving functions, *i.e.*, breaking, accelerating and steering.

4) Decision making, path planning and following while respecting the regulations of traffic.

5) Information collection and exchange, such as maps, traffic status and road incidences.

6) Platooning with other vehicles.

The work presented in this paper is build on the authors previous work [2] which considers a visual line guided system to control the steering of on-board Autonomous Guided Vehicle (AGV) *that pursue a guided path.* This work was initiated with a project done in collaboration with Siemens Spain S.A. focusing on the development of a driver assistance system for buses in the city center using a vision-based line guided system. The main idea was that the driver should still be able to actuate the brake pedal in order to avoid any potential collisions, but has no longer a steering wheel to manually guide the vehicle. Following the previously mentioned level of autonomy, the system presented in this paper could be assigned to a level somewhere between the levels of partial or high autonomy. The presented control system approach takes over the complete control of the steering wheel which corresponds to a high level of autonomy. However, the speed is controlled as an assistance cruise control by keeping the user's desired speed under the limitation of the maximum speed in each section of the predefined path even if the user pushes the gas pedal to exceed this limit. The speed assistance control also allows to stop the vehicle in case of an emergency, such as the detection of an absence of the line, the push of an emergency button or if a specific localization mark on the road occurs.

The localization of the vehicle on the predefined path was implemented with the help of specific visual localization marks. In cases of a false or missing detection of one or more localization marks, the localization is supported by an odometry approach, i.e. the integration of the speed of the vehicle. These marks were not only used to localize the vehicle, but also to provide additional information to the control system and to the assistance cruise control. This additional information noticeably improves the behavior of the system, such as allowing to reach higher speeds and improving the robustness of the system. Regarding the previous list of the capabilities of a full autonomous system, we are here focusing on Point 1 and partially Point 3. The collision avoidance control is out of scope of this work. Various and potentially adverse conditions of the road such as on rainy days are also not considered in the work at hand. Therefore, the main focus is to find a low cost solution for a vision-based control approach including (I) the steering of an autonomous vehicle using a line guide, (II) a speed control assistance and (III) the localization of the vehicle, while preserving robustness against brightness variations in inner-city environments.

The layout of the paper is as follows. Section 2 discusses related works on autonomous cars. Section 3 describes the full system architecture as well as the low-level car controller (Section 3.2), and the human-machine interface developed to remotely supervise the system, to command the desired speed and to realize an emergency stop if needed (Section 3.3).Section 4 presents the derived computer vision algorithms. Section 5 presents the general system architecture, the car automation and the control approach of the steering wheel. The results of experiments that were carried out in a closed test road are presented in Section 6. Finally, Section 7 presents the conclusion and the future work of this paper.

2. RELATED WORKS

Autonomous guided vehicles (AGVs) are generally used in manufacturing and logistic systems inside warehouses, but their acceptance inspired many other applications such as guided buses in city transportation. They were introduced during 1950s and, by 1960s the Personal Rapid Transit (PRT) began to advent. Different guidance systems were introduced for AGVs such as systems based on optical distance measurements, wires, magnetic tapes or computer vision. Each type is based on own design requirements and comes with own related advantages and disadvantages. For instance, in wire guidance systems, a wire is installed below the floor on which the AGV is moving. The wire emits a radio signal which can be detected by a sensor on the bottom of AGV close to the ground. The relative position to the radio signal is applied by the AGV to follow the path. In magnetic tape guidance system, a flexible tape of magnetic material is buried in the floor or road like in the case of the wire guidance system. The advantage of this method with respect to the wire guidance system is the fact that it remains unpowered or passive. In laser navigation systems, the AGV is equipped with a laser transmitter and receives the reflection of the laser from a reflective tape installed on the surrounding walls. The localization and navigation is done using the measurements of the angles and distances to the reflectors. However, this method is generally only used in indoor environments [3,4]. A vision based navigation system uses a vision sensor to track landmarks in the environment which means that no magnets, no induction wires and also no laser technique is required to let the AGV follow a specified path [5,6].

On the other hand, the motivation to reduce traffic jams, to improve the fuel economy and to reduce the number of vehicle accidents in transportation leads to the introduction of different levels of automated driving. Many research institutes and automotive manufacturers worldwide are introducing their automated driving solutions, based on proprioceptive sensors such as the Anti-lock Brake System or the Electric Stability Program, or based on exteroceptive sensors such as radar, video, or LiDAR [7]. The very first experiments on autonomous vehicles have been started in 1920s1920s and promising steps have been conducted in the 1950s. The research in autonomous driving in Europe started within the PROMETHEUS project (program for a European Traffic with Highest Efficiency and Unprecedented Safety) which was one of the largest research projects in fully automated driving in 1986. The obtained results of this project are regarded as milestones in the history of vehicular and robotic systems. Two of the vehicles were ARGO by VisLab [8,9] and VaMoRs [10] tested in 1998. Both of them used two cameras to detect road lanes and to avoid obstacles, but the implemented algorithms and strategies were different. In 1995 the NAHSC project (National Automated Highway System Consortium) started in the United States within the California PATH (Partners for Advanced Transit and Highways) program [11]. In 1997, the important Demo'97 was developed in San Diego in which some cars were guided by a magnetic guided line inside the asphalt. An array of different sensors had been installed in those cars to execute self-driving tests and to form automated platoons of 8 cars.

In the last decade many authorities around the world introduced plans to promote the development and establishment of automated vehicles [12].

Numerous commercial vehicles offer some levels of automation, such as adaptive cruise control, collision avoidance, parallel parking system, lane keeping assistance, *etc*. Research on this topic got a strong impulse by the challenging test-bed of DARPA in the grand and the urban challenge in 2005 and 2007 [13] with impressive results obtained by Sebastian Thrun and his team from the Stanford University in 2005 [14] and 2008 [15], or by the Braunschweig University in 2009 [16]. All of these works tried to cover all the capabilities listed for a fully autonomous system, which is also the case for the recent Google Car [17]. In this specific case, the obtained results of this approach should enforce legal changes to achieve the first license for a self-driving car. The European Union also has a long history of contributing to automated driving such as the Vehicle and Road Automation (VRA) program, the GCDC (Grand Cooperative Driving Challenge), and others. Many countries plan to develop sensors, control systems and services in order to have competitive autonomous driving systems and infrastructures [18]. A considerable number of studies and projects have been funded or are still continuing within the new HORIZON2020 research framework in the European Union.

For instance, a Mercedes-Benz S-Class vehicle equipped with six radar sensors covering the full 360° angular range of the environment around the vehicle in the near and far range has been introduced in 2013. The vehicle drove completely autonomous for about 100 km from Mannheim to Pforzheim, Germany, in normal traffic [19].

Moreover, there are also some works focusing on the sixth point of the list of autonomous capabilities (*i.e.*, the platoon formation). In 2010, the multidisciplinary European project SARTRE used new approaches in platoon formations and leader systems to successfully present an autonomous platooning demo traveling 120 miles [20]. The platoon comprised one human-driven truck followed by four cars equipped with cameras, laser sensors, radar and GPS technology. A complete test of different systems of leader following, lane and obstacle detection and terrain mapping has been done by the VisLab. In 2010, the laboratory directed by Alberto Broggi covered the distance of 15.92615.926 km from Parma to Shanghai with a convoy of four motor homes [21,22]. All of them were equipped with five cameras and four laser scanners, no road maps were used. The first vehicle drove autonomously in selected sections of the trip while the other vehicles were 100%100% autonomous, using the sensors and the GPS way-points sent by the leader vehicle. The control of speed and/or steering of autonomous vehicles with a localization system based on GPS information is also presented in the literature, see, e.g., [23]. Herein, a cruise control approach for an urban environment comprising the control of the longitudinal speed based on the speed limits, the curvature of the lane and the state of the next traffic light is proposed. In [24], control tests of a high-speed car running the Pikes Peaks rally drive are presented. The work in [25] shows a localization system without GPS, based on the detection of intersections and the use of a virtual cylindrical scanner (VCS) to adapt the vehicle speed.

Highly automated levels of driving require a very wide range of capabilities like sensing the environment, figuring out the situation and taking proper action for the driver. The design of a cost-effective solution for such highly automated driving systems is challenging and most of the time leads to an expensive multi-sensor configuration like the way introduced in [26]. Vision-based systems are considered to be a cost-effective approach for automated driving systems [27]. Vision-based systems can be categorized in different research areas and applications in the field of automated driving such as distance estimation using stereo vision [28,29] or monocular vision data [30]. A review of the literature in on-road vision-based vehicle detection, tracking, and behavior understanding is provided in [31]. From an algorithmic point of view, computationally more complex algorithms require an understanding of the trade-off between computational performance (speed and power consumption) and accuracy [6,32]. For instance, an offline-online strategy has been introduced in [33] to overcome this trade-off. Furthermore, vision-based systems have many applications in automated driving, such as road detection which is one of the key issues of scene understanding for Advanced Driving Assistance Systems (ADAS). In [34] road geometries for road detection are classified, and [35] introduces an improved road detection algorithm that provides a pixel-level confidence map. The paper [36] describes a neural network road and intersection detection. Another vision-based application for ADAS is lane keeping assistance, where a technique for the identification of the unwanted lane departure of a traveling vehicle on a road is presented in [37].

Despite some new and improved computer vision algorithms which have been introduced in recent years such as [38], it has to be noted that the variation in the lighting conditions, occlusions of the lane marking or road shoulders, and effects of shadows make the current vision-based solutions not reliable for the steering control of an autonomous car. Furthermore, these algorithms are still not completely real-time capable to be used in the closed control loop. Based on that and the specific constraints of our project mentioned in the previous section, this work focuses on a vision-based line guided system approach. To the author's best knowledge, a vision-based line-guided system has not been presented to control the steering of an autonomous car under the maximum speed constraints of urban environments.

3. SYSTEM DESCRIPTION

In this section the system description of the automated car is presented, which can be divided in the general system architecture, the vehicle automation and the human-machine interface.

3.1. General System Architecture

The system architecture comprises several components as depicted in Figure 1. The first component is the computer vision system (described in Section 4), also called "Visual processing", which generates the information of the local position of the vehicle with respect to the path to follow (painted as a line). In addition it is able to detect binary visual marks painted on the road, which have the encoded information of its global positioning on the track as well as the maximum speed inside this section and the curvature radius. This information is stored in a data base.

The lateral control system described in Section 5 includes the "guidance controller". These are two feedback controllers which have to keep the vehicle on the path by minimizing the deviation from the path. This component also includes the "Steering Offset".

A "positioning" system allows to calculate the position of the vehicle on the track. The odometry information is generated by integrating the vehicle's speed and updated with the absolute information obtained by reading the visual marks. In case a mark is not detected at the begining of one section, the odometry position estimation is used to change from one section of the track to another.

The "speed decision" system limits the speed of the vehicle by taking its position on the circuit into account. Herein the human driver is not permitted to overwrite this information either manually or using the Human-Machine Interface (HMI, see Section 3.3).

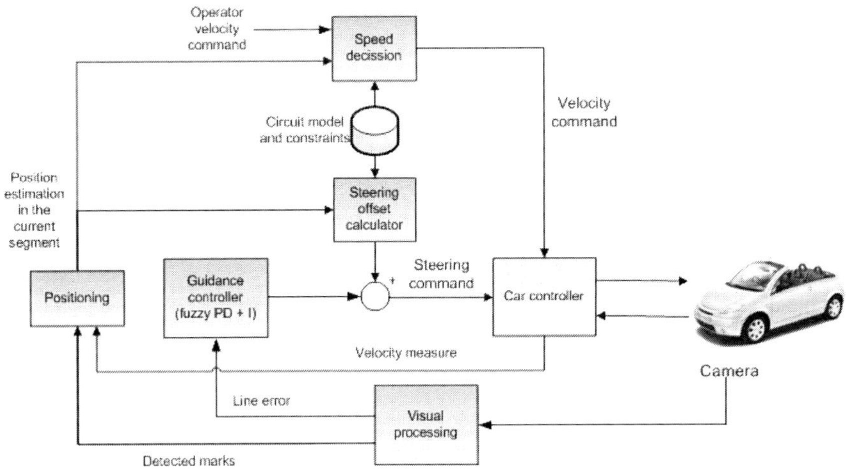

Figure 1. Full system architecture.

Both velocity and turning commands are sent to the "car controller" system (see Section 3.2), which actuates directly with the motors installed in the vehicle and reads the measured speed of the vehicle.

3.2. Vehicle Automation: Car Controller System

Within the project considered in this paper, a passenger car (shown in Figure 2a) has been fully automated. A detailed description of the vehicle automation is included in [39] and the steering and speed fuzzy controllers were described and tested in [40]. The automation includes a control unit that communicates with a data acquisition card that provides signals to the electronic accelerator pedal and to a servo amplifier operating on the vehicle steering assistance motor. This solution is applicable in case of vehicles with an electric steering assistance system. In the case of hydraulic assistance, solutions as shown in [41] could be used. Moreover, the control unit sends commands to the electric motor controller which operates the brake pedal. As a safety measure, a remote control could operate on this motor so that any control signal is blocked and an emergency braking signal is provided from the power source. The fuzzy controllers allow, firstly, a vehicle behavior similar to that of a human driver. In addition, functions such as emergency manoeuvres (emergency brake and steering avoidance manoeuvres respecting conditions of stability) which are generated if an obstacle is detected on the route have been implemented in the decision module [39]. Figure 2b shows the assembly of the electronic control units of the automated vehicle. Moreover, this low-level control layer receives the desired signals of steering wheel angle and vehicle speed from a high-level control layer and Figure 2c shows an overview of the architecture of the internal automatic control of the vehicle.

This architecture allows the implementation of different types of applications such as a GPS-trajectory guided vehicle, line tracking, collision avoidance applications, remote control from a mobile device like a smartphone, *etc*.

3.3. Human-Machine Interface for the Driver Supervision

In this work, an autonomous driving system is presented which is always under the supervision of a driver, either a remote external operator or a driver inside the vehicle. For that reason the presented HMI for the supervision task was designed, see Figure 3. It shows the current information of the vehicle ("*vehicle info*"), comprising the desired user speed ("*user speed*") that can be changed in real time, the current desired speed ("*last speed command*"), the current speed ("*current speed*"), the maximum speed reached during all the experiment ("*maximum speed*"), and the distance traveled in the current section of the track ("*distance traveled*"). Below this information, the information related to the current section of the track in which the vehicle is traveling ("*track section info*") is shown. The first line shows the id of the current section ("*current section*"), followed below by the curvature radius of the current section of the track ("*curvature radius*"), the speed limit inside the section ("*speed limit*"), and the limit for the transition between the previous section and this section ("*transition speed limit*"). Any system messages appear at the bottom of the HMI as a state of the system ("*System status*"), e.g., a message indicating a breakdown of the communication or any camera error is shown in this field. On the right side of this bottom part of the HMI, an emergency stop button (yellow and red button) was also installed. Once this button is pushed, the vehicle will stop whatever its state is. A real-time image feedback of the camera is also shown in the HMI. Over the image feedback, the id of the last mark detected ("*last code detected*"), the state of the camera, and control communications ("*communication status: video, control*") is shown, respectively. Both communications are managed with a specific software daemon to reconnect autonomously after the detection of a communication breakdown.

(a)

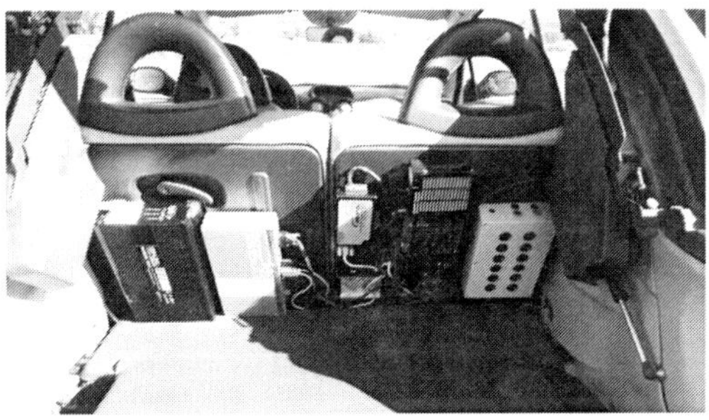

(b)

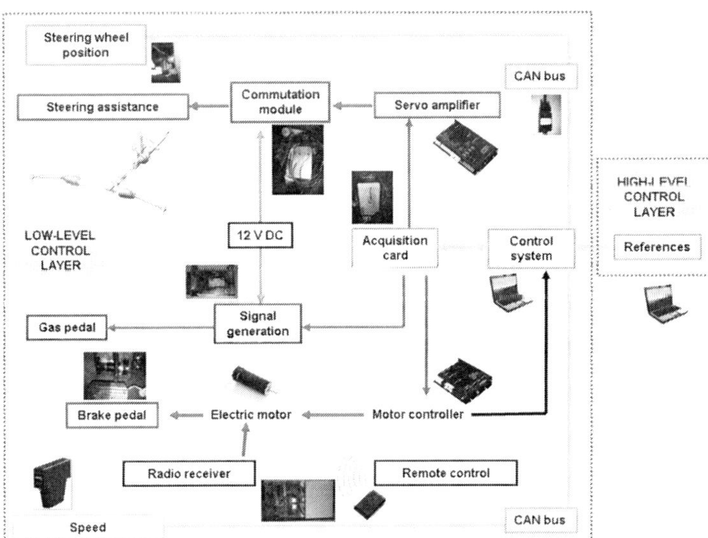

(c)

Figure 2. Passenger vehicle automation. (**a**) Autonomous vehicle; (**b**) electronic deployment of the control system; (**c**) internal architecture for vehicle control.

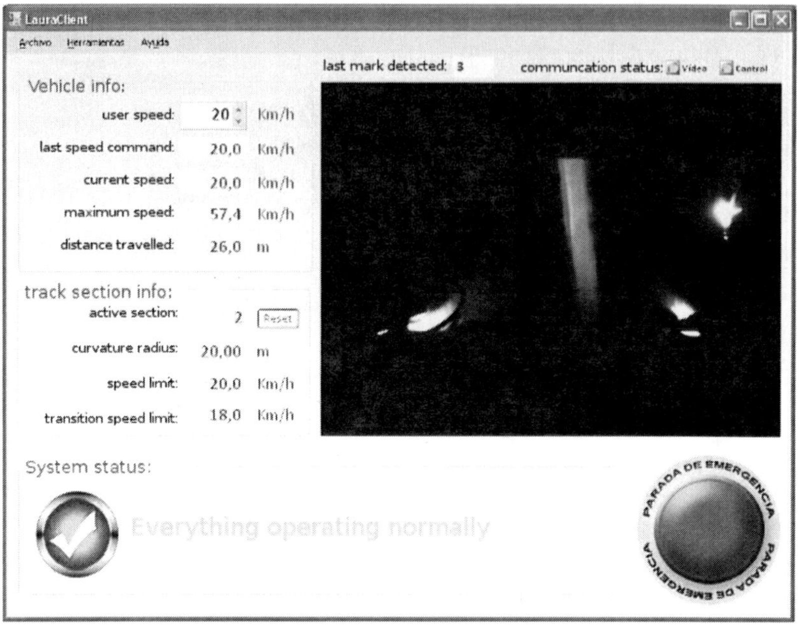

Figure 3. Developed human-machine interface.

4. COMPUTER VISION SYSTEM

A computer vision algorithm processes in real time the images captured by a monocular camera under illumination with ultraviolet (UV) light. This camera is placed in the front part of the vehicle, looking downwards and isolated from the sunlight by a black box. A similar approach with a looking downward camera at the bottom of the car is presented in [42]. In this work the car structure was used to avoid that the illumination changes affect the image acquisition. The camera was used to get information from the road to do a localization matching between vision based ground features and the global localization done with a RKT-GPS system. No control of the steering wheel was presented in this work. We, in the presented work, have to set the camera on the front of the vehicle since we are controlling the steering wheel in an Ackermann model vehicle. The mentioned approach could be used in our case to get the information from the visual marks painted on the road but not to guide the vehicle.

The presented algorithm detects both the line to be followed by the vehicle, as well as visual marks painted on the road. The visual marks provide a coded information associated to forward path properties like curvature, maximum allowed speed, *etc.*, which is used by the controller to anticipate changes and react faster.

Two kinds of paint were used for the line and the visual marks. Due to their different pigments, the line is seen as blue while the marks seen as yellow in the images captured by the camera under illumination with UV light. The rest of the image remains black, as depicted in Figure 4.

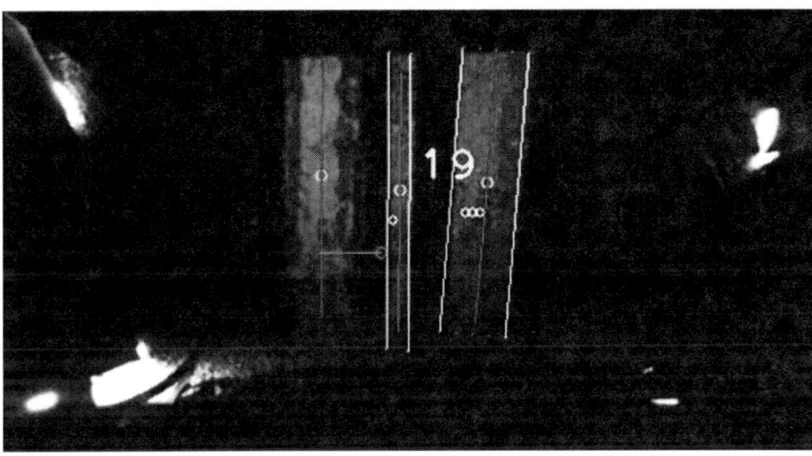

Figure 4. Line and mark detection by the computer vision algorithm.

The visual algorithm has been designed with a special focus on robustness and thus being able to detect fragmented lines due to occlusions or small irregularities on the road. The full system has been tested under different weather conditions including sunny and cloudy days as well as sparkling days. To evaluate the robustness of the system it is important to know the exactly speed at which the system is able to see every single cm of the track by the camera installed. To know this value it has to be taken into account that the frame rate of the camera is equal to 29 fps and the size of the camera system (the metallic box) is 30×5030×50 cm. The distance covered by the system is equal to 29 fps×30×30 cm=870=870 cm/s, and 870 cm/s are equal to 31.3231.32 km/h. That means that at this speed the captured frames cover all the track without losing any single cm of the road. When the speed of the vehicle is higher than 31.3231.32 km/h the system will lose some distance covered by the vehicle in between each frame captured. In the case of 40 km/h, which is equal to 1131 cm/s, the system cover a distance of 39 cm per frame, and for 50 km/h (1388.81388.8 cm/s) the system cover a distance of 47 cm per frame. That means that the system can not see 9 cm and 17 cm in between each frame for the speed of 40 and 50 km/h respectively. Despite this limitation of the vision system the vehicle was able to detect the line and visual marks covering successfully long distance at different speeds. The computer vision algorithm has two different parts which will be described in the following sections.

4.1. Line Detection

This first part of the visual algorithm processes the current acquired image to obtain information about the line to be followed by the vehicle. If there is a line in the analyzed image, the line angle and distance are determined with respect to the image center.

The first step for the line detection is color segmentation on YUV space that exploits the blue appearance of the line in the image. Some other color spaces were tested, but YUV provided better results under different light conditions. A rectangular prism inside the YUV space is defined so that only the pixel values inside this volume are considered part of the line. The output of this first step is a binary image in which only the line pixels are set. This method proved to be robust in detecting lines of different blue tones and brightnesses.

To reduced the noise, a second step is performed. In the binary image, every 8-connected pixel group is marked as a blob. Blobs having an area outside a defined range are discarded. Then, for every survivor the centroid, the dominant direction and the maximal length are computed. Those blobs with a too short maximal length are ignored. The remaining blobs are clustered according to proximity and parallelism, so each cluster becomes a candidate line. The centroid and dominant direction of each candidate line are calculated from the weighted sum of the features of its component blobs, where the weight of each blob is proportional to its relative area. In this way the algorithm can accurately detect lines that are fragmented because of the aging of the paint.

The last step consists of the choice of the winning detected line from the whole set of candidate lines. The decision is achieved by using temporal information between the current and the previous frame, *i.e.*, the candidate closer to the last frame winner in terms of centroid distances will be selected as the current frame winner. This rejects false positives because of old line traces along the circuit. In the case that all candidates are far enough from the last frame winner, a bifurcation is assumed and the winner will be the leftmost or rightmost candidate, depending on the information associated to the last detected visual mark.

4.2. Mark Detection and Decoding

The second part of the computer vision algorithm includes the detection and decoding of visual marks painted on the road next to the line to follow. The visual marks are detected and decoded even when they appear rotated in the image as a result of vehicle turns.

Each visual mark is labeled through a unique identifier that represents a binary-encoded number, where mark bits are drawn as bars parallel to the line. Because of the reduced visual field of the camera, instead of painting a bar for each bit like in common barcodes where the width of the bar depends on the bit's value, a more compact encoding scheme was chosen. All bits are bars with the same width with no spacing between them. When a bit is one, the bar is painted; when it is zero, the bar space is left unpainted. In the image a mark appears as a set of yellow rectangles. Every painted rectangle is a bit with value one.

A start bar is added at the furthest bit slot from the line to designate the beginning of the mark. The mark specification also defines the number of bits per mark, a fixed bit width, a minimum bit length, and valid ranges for line-mark angle and line-to-start-bit distance. According to the specification, the algorithm will only detect marks that are placed on the right of the line in the direction of motion of the vehicle.

Similarly to the line detection phase, the mark detection algorithm follows a set of steps. First of all, the acquired image is segmented by color in YUV space, extracting the potential mark pixels. The color space boundaries of this segmentation are set so that it admits several yellow tones, ranging from tangerine yellow to bright green, as seen in tests with multiple paints. This makes the color-based filter less restrictive and avoids false negatives.

As the probability that any yellow noise present in the image has a valid mark structure is low, the following steps of the algorithm seek for this valid structure in the color-segmented pixels to reduce the false positives.

After the color segmentation, the resulting eight-connected pixels are grouped in blobs. The blobs that do not meet the following criteria are considered as noise and are discarded: the blobs must appear at the right of the line, the blob area must be in a valid range (computed from the visual mark specification), the angular distance between the dominant blob direction and the line must be in the specified range, and the blob length in the dominant direction must be larger than the minimum bit length.

The blobs that pass the filters correspond to a set of bits with value 1. A pattern matching determines the specific ordinal number of bits for each blob. Assuming each mark has a total of *N* bits (including the start bit), *N* pattern tests are carried out for each blob, one test for each bit in the range [1,N]. For every bit *i*, the pattern Pi,BP is a rectangle with the direction and length of the blob *B* and a width *a* equal to *i* times the bit width in the specification (in pixels). The ordinal number of bits for the blob *B* will be the value *i* that minimizes the cost function present in the Equation (1).

$$c(i,B) = f(i,B) + \hat{g}(i,B) \quad (1)$$

where

$$f(i,B) = 1 - \frac{a(B \cap P_{i,B})}{a(B)}$$
$$\hat{g}(i,B) = \frac{g(i,B)}{g_{max}}$$
$$g(i,B) = | a(B) - a(P_{i,B}) |$$

and a(S)a(S) is the area in pixels inside the shape *S*.

Function *f* indicates how much the pattern covers the blob. Function *g* evaluates the similarity between the blob and pattern areas. Patterns whose *f* or *g* are above a threshold are discarded. This forces the best solution to have a minimum quality in both indicators. Then, the minimization process favors patterns that cover the blob while having a similar size. The threshold for *g* is gmaxgmax and g^g^ is a normalized version that stays in [0,1][0,1], like *f* does.

After assigning a number of ones to all processed blobs, the rightmost one is interpreted as the start bit. If its distance to the line is in the range allowed by the specification, the mark remains in the detection process, otherwise it is ignored. The mark's dominant direction is computed as the average of all its blobs. The orthogonal vector to this direction defines a baseline that is divided into *N* consecutive fixed-width bit slots, starting from the start bit. All bit slots are set to an initial value of zero. Then, blob centroids are projected on the baseline and each projection falls into one of the slots, an then filled with a one. Adjacent slots are also filled with ones according to the blob's number of bits. Finally, the slot values define the binary-encoded mark identifier whose least significant bit is the closest to the start bit.

The final detection step of the visual mark identifier that is going to be passed to the control system is elected with a two different voting process. The first voting process evaluates the detected visual mark id in each frame. Working under the assumption that the code is always located on the right of the line, this part of the image is divided in nine horizontal sections. In each of these section the system tries to identify a visual mark. The final result comes from the most detected code. An example of this voting process is shown in the Figure 5.

Figure 5. Single frame voting process for the detection of visual marks.

The second voting process evaluates the code detection among the last *M* frames. Large values of *M* produce higher detection delays but increase detection robustness, as more image samples are taken into account. In our experiments, M=3M=3 gave good results. Besides the detected mark identifier, the algorithm provides an estimation of the number of frames and time lapse since the mark was last seen. This information is especially useful at high speeds and high values of*M*, when the decision is delayed until *M* frames have been captured, but the mark was only seen on the first few frames. In addition, an estimation of the mark quality is given based on its average luminance. Figure 4 shows the detection of the line and the mark which represent the number 19.

Once the mark detection stage provides a visual mark identifier, the decoding stage starts. The information encoded in these marks is the current location of the marks on the track, the size and the curvature of the following section of the track, and the maximum permitted speed on it.

The detection of a mark is always checked with a database of available marks, avoiding false detections that could localize the vehicle in another section of the track.

5. LATERAL CONTROL SYSTEM: A VISION-BASED FUZZY-LOGIC CONTROLLER

The steering control system of the vehicle includes three additive components: The first one is a fuzzy-logic feedback controller that acts with a behavior equivalent to a PD controller. The second one is the weighted integral of the error (distance between the line reference and the measured line position). The third component is the steering offset that acts like a feedforward controller, by changing the operating point of the fuzzy-logic controller to improve its performance, based on the track information given by the detected mark. All the three components are added at the end of the control loop to generate the output of the control system, making a structure of Fuzzy+I+OffsetFuzzy+I+Offset, as shown in Figure 6.

The objective of this work is to develop a controller for a track with small radius curves. In such conditions the speed of the car is not allowed to be very high (less than 50 km/h).

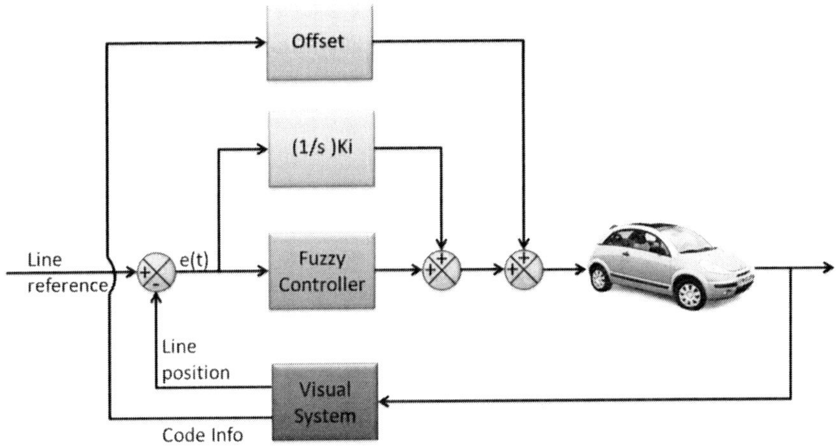

Figure 6. Simplified control loop of the visual control system.

From several real experiments with the vehicle, the authors can confirm that it is practically impossible for a human pilot using just the information received from the down-looking camera to drive faster than 10 km/h while keeping the line-following error low enough to meet the requirements of the application. This is because the pilot only sees 39 cm ahead, and, at that speed, the contents of this area change completely every 0.1080.108 s.

The first and main component, the fuzzy-logic feedback controller was implemented using a software called MOFS (*Miguel Olivares' Fuzzy Software*). This C++ library has been previously successfully used to implement fuzzy-logic control systems with other kind of robotic platforms such as an unmanned helicopter for autonomous landing [43] or quadrotors for avoiding collisions [44]. Thanks to this software, a fuzzy controller can be defined by specifying the desired number of inputs, the type of membership functions, the defuzzification model and the inference operator. In [43], a more detailed explanation of this software is provided.

The fuzzy-logic controller designed with PD-like behavior has two inputs and one output. Triangular membership functions are used for the inputs and the output. The first input is called the error and is the difference between the line reference and the measured line position in pixels, with respect to the center of the image (Figure 7a). The second input is the derivative of this error (Figure 7b). The output of the controller is the absolute turn of the steering wheel in degrees to correct this error (Figure 7c).

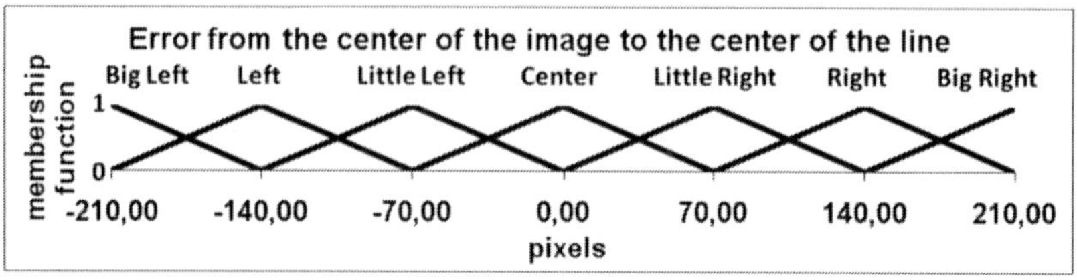

(a)

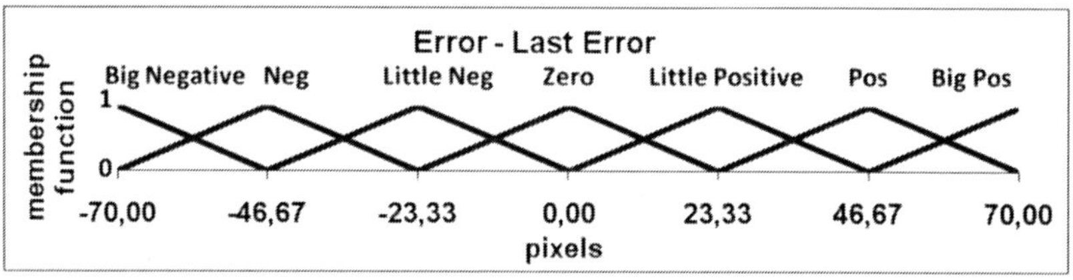

(b)

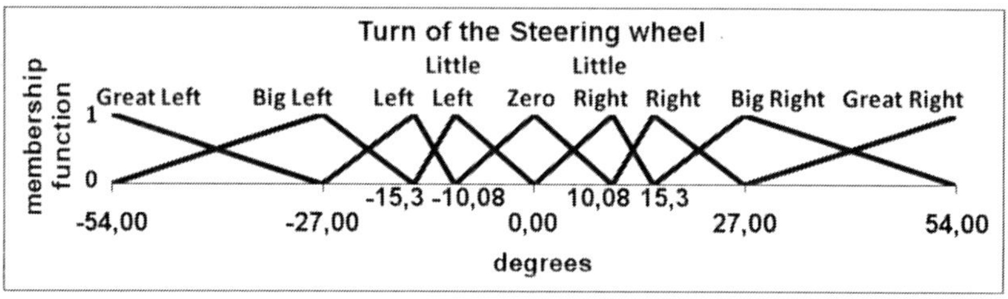

(c)

Figure 7. Definition of the two inputs and the output of the fuzzy controller. (**a**) First input variable of the fuzzy controller: Difference between the line reference and the measured line position with respect to the center of the image, in pixels; (**b**) second input variable of the fuzzy controller: The difference between the last error and the actual, in pixels; (**c**) Output variable of the fuzzy controller: The steering wheel angle, in degrees.

The rule base of the presented fuzzy control component is formed by 49 if-then rules. Heuristic information has been used to define the output of each rule as well as for the definition of the range and set of each variable. The developed fuzzy system is a Mamdani-type controller that uses a height defuzzification model with the product inference model as described in Equation (2).

$$y = \frac{\sum_{l=1}^{M} \overline{y}^{l} \prod_{i=1}^{N} \left(\mu_{x_i^l}(x_i)\right)}{\sum_{l=1}^{M} \prod_{i=1}^{N} \left(\mu_{x_i^l}(x_i)\right)} \tag{2}$$

Herein *N* and *M* represent the number of input variables and the total number of rules, respectively. µxli denotes the membership function of the *l*-th rule for the *i*-th input variable. y¯l represent the output of the *l*-th rule.

To tune the fuzzy-logic controller, a driving session performed by a human driver at 10 km/h provided the necessary training data to modify the initial base of rules of the controller and the size of the fuzzy sets of its variables. For the definition of the initial fuzzy sets, a heuristic method was used based on the extraction of statistical measures from the training data. For the initial base of rules, a supervised learning algorithm implemented in MOFS has been used. This algorithm evaluates the situation (value of input variables) and looks for the rules that are involved in it (active rules). Then, according to the steering command given by the human driver, the weights of these rules are changed. Each time the output of an active rule coincides with the human command its weight will be increased. Otherwise, when the output differs from the human command its weight will be decreased by a constant. Anytime the weight of a rule becomes negative the system sets the output of the rule to the one given by the human driver.

Since the velocity of the vehicle is not included in the fuzzy controller, this is taken into account by multiplying the output of the fuzzy-logic controller by 10v10v, where *v* is the current velocity of the vehicle. The definition of the numerator value of this factor is based on the velocity, in km/h, obtained during a driving session with a skilled human driver, in which data was acquired to tune the rule base of the fuzzy controller.

The second component of the presented control system is the weighted integral of the error. The objective of this component is to ensure that the error converges to zero in every kind of track. The output of this component follows Equation (3).

$$I_t = I_{t-1} + e \times \frac{1}{t} \times Ki \tag{3}$$

Herein *e* is the current error between the center of the line and the center of the image, *t* is the frame rate, and KiKi is a constant that appropriately weights the effect of the integrator. In the presented control approach this constant KiKi has a value equal to 0.60.6.

Finally, the third component of the lateral control system is a steering offset component. It behaves like a feedforward controller that offsets the effect of the change of the curvature of the circuit in every different track, updated each time that a new mark has been detected. It is theoretically calculated using the equations of the Frenet-frame kinematic model of a car-like mobile robot [45]. More detailed information about this control component can be found in [46].

6. EXPERIMENTS

All the experiments presented in this work were done inside a closed circuit in the infrastructure of INSIA-UPM at the University Institute of Automobile Research, Technical University of Madrid. Figure 8 shows the design of the track. The track has two curves of 11 and 20 m of radius and two straight lines connecting the two curves. A access straight line with a length of 100 m was design to go inside the track and test higher speeds. Each lap of the track has a length of 245 m.

A set of different type of experiments were defined to evaluate the behavior of the full system. Firstly, the experiments related to the computer vision system are presented. Secondly, the different tests to evaluate the autonomous driving and the speed assistance systems are shown.

6.1. Computer Vision System

The robustness of the computer vision system was tested in different kind of experiments. Firstly, is presented the robustness of the system against different type of line occlusions. Secondly is presented the behavior of the system against visual marks occlusions and false detections. For each different experiments a set of capture frames are shown, but it must be taken into account that not all the processed images could be saved and are not shown. This is caused by the computational cost of saving images working at the same time as the image processing algorithm. All the experiments presented in this subsection were tested for 13 laps at the speed of 15 km/h and for six laps at different speed from 14 to 42 km/h. During all these experiments the car was drove autonomously. All the occlusions, fork lines, and false detection were overpassed one time for each lap. A slight increase of the RMSE was noticed with respect to similar tests without the occlusions. However, the vehicle complete successfully the 13 at constant speed and the six laps with variable speed without loosing the control in any moment. A comparison table of all the experiments done included the presented on this subsection are shown in the Subsection 6.2.

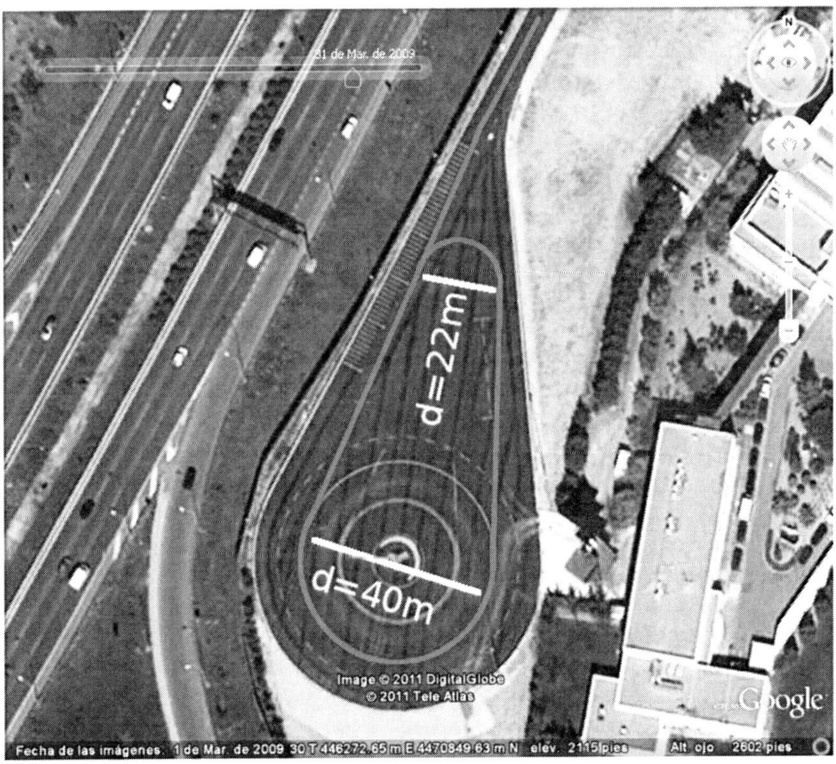

Figure 8. Representation of the circuit at INSIA installations taken with the Google Earth tool.

6.1.1. Line Detection Robustness

The robustness of the vision algorithm to detect the guided line was tested in different situations. Here, it is presented the behavior of the system against a total occlusion of the line, reduction of the width of the line, and presence of other lines.

6.1.1.1 Total Occlusions

A total occlusion is defined as the complete occlusion of the line for a certain distance. The robustness of the system against total occlusion was done to evaluate how much time the system could drive autonomously in blind conditions due to the non-detection of the line. This could be caused by external elements or the complete degradation of the paint of the line itself. Different sizes of occlusions were installed inside the circuit on straight lines and curves.

As was explained in Section 4 due to the limitations of the computer vision system the algorithm lost a maximum of 17 cm when the vehicle is driving at the maximum speed of 50 km/h. Here a higher occlusions equal to 30 and 50 cm inside a straight line were tested. Depending on the vehicle speed and when the frames are captured the line can be completely covered and can not be detected for at least one frame in the case of the 30 and 50 cm occlusions. In the presented experiment for a occlusion of 30 cm at 15 km/h a small section of the line was detected in all the frames. In Figure 9 it is shown all the frames captured by the computer vision system during the overpassing of this occlusion.

In the case of 50 cm inside straight line the computer vision system lost completely the line for at least one frame, and the occlusion was saw for at least five frames. The Figure 10 shows the images captures during the execution of this test.

The behavior of the system for line occlusions inside curves was also tested. In this case the system was successfully tested for occlusions of 10 and 30 cm. Higher distance occlusions in curve affect the system behavior and could not be 100%100%certified that the system could operate in satisfactory conditions. Figure 11 shows the images captured for the test of the 10 cm occlusion and Figure 12 shows the images captured during the test of line occlusion for 30 cm.

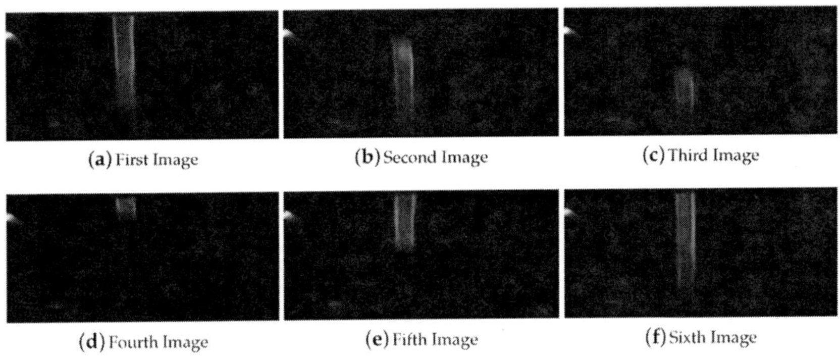

Figure 9. Sequence of images for the line occlusion for 30 cm inside a straight line.

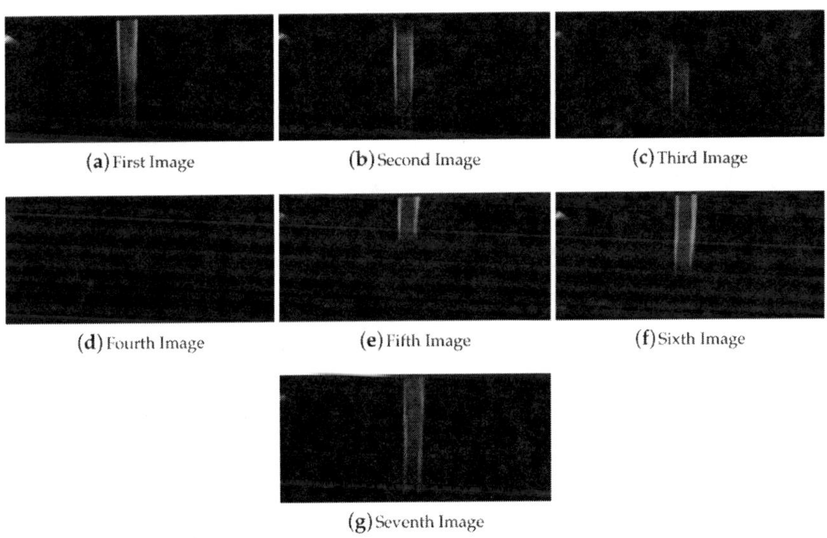

Figure 10. Sequence of images for the line occlusion for 50 cm inside a straight line.

Both occlusions are located consecutively. The last frame in Figure 12 is the precedent image of the first one in Figure 11.

Higher occlusions than 50 cm inside straight lines and 30 cm inside curves were tested without 100%100% of success, being them handled by the system only in some specific situations. Based on the presented tests a safety measurement was included to do an emergency stop in case the line was lost for 1 m. This value could be increase or decrease based on the safety requirements of the final system.

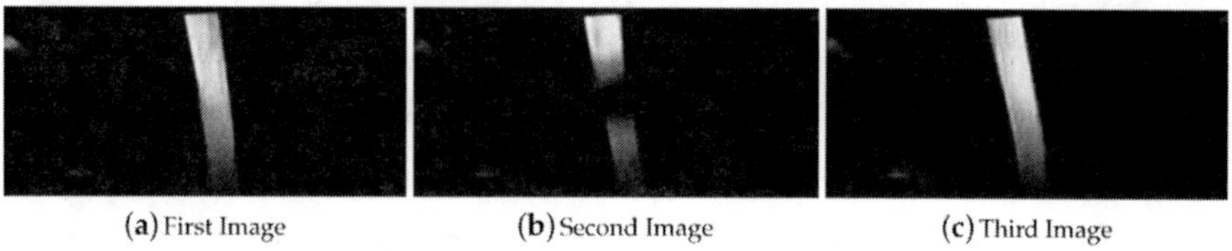

Figure 11. Sequence of images for the line occlusion for 10 cm inside a curve.

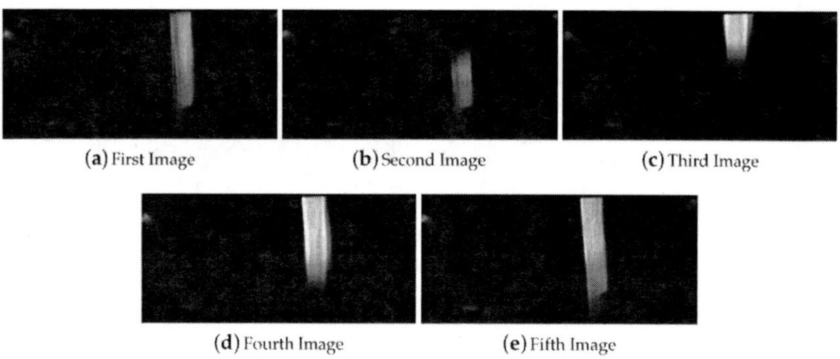

Figure 12. Sequence of images for the line occlusion for 30 cm inside a curve.

Figure 13 shows the evolution of the system when all the mentioned occlusions were included in the track for a constant speed equal to 15 km/h. As is shown in this Figure, the autonomous lateral control system managed to command the vehicle without loosing the line and only increasing slightly the total RMSE value to 5.80985.8098 cm in comparison to other similar tests without occlusions, which are presented in Section 6.2.

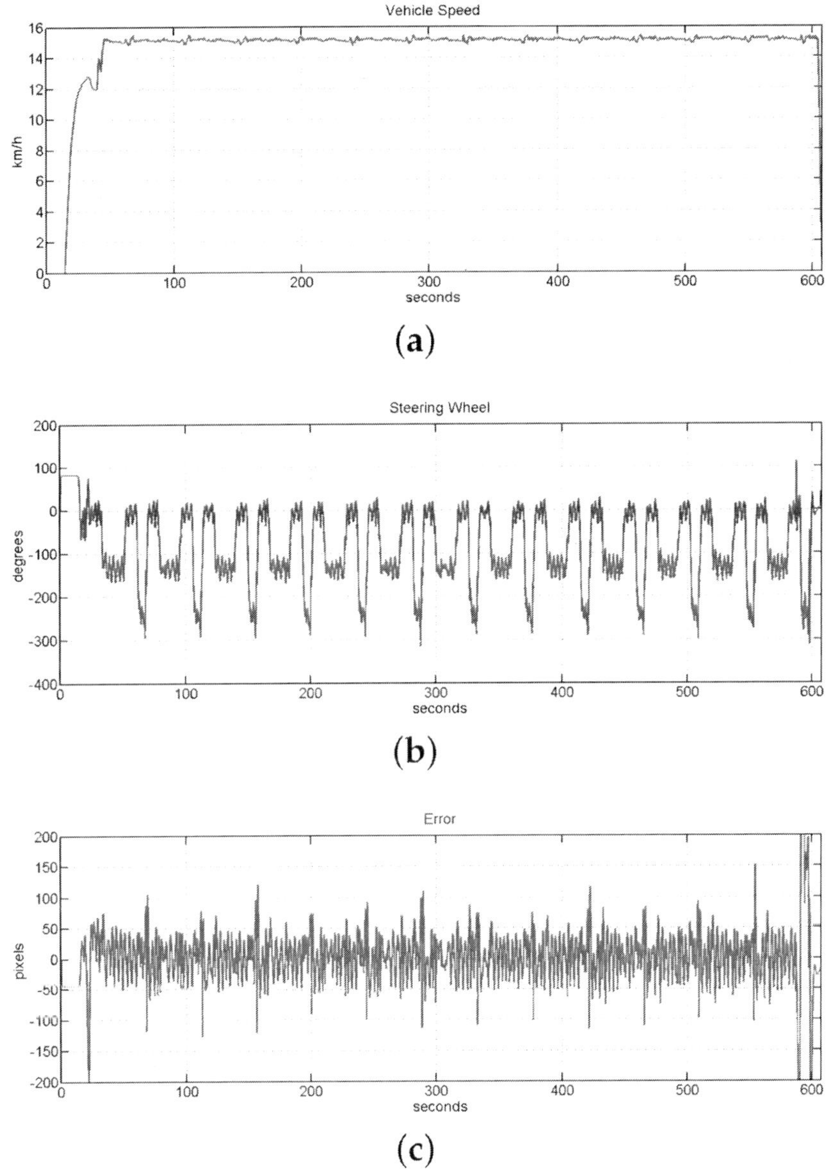

Figure 13. Evolution of the system during 13 laps inside the track with mark detection and different line occlusion onside straight line and curves. Speed was set to 15 km/h. (**a**) Vehicle speed evolution; (**b**) evolution of position of the steering wheel; (**c**) evolution of the error measured in pixels, the value of RMSE for this test is equal to 5.80985.8098 cm.

6.1.1.2 Line Width Reduction

In this experiment, the line width was set based on the available equipments to paint the line on the road. It was painted manually and the width was in between 5 to 6 cm. Here is presented the experiments done to check the robustness of the system facing thinner lines. The reduction of the width of the line could be caused by a partial occlusion of it, the degradation of the line or the presence of paddles. Furthermore, another objective of these experiment is to evaluate the system behavior with different width of the line for a future real application of the system. A reduction of the 25%25% and 50%50%of the line width were applied in a section were the width was equal to 5 cm. Figure 14 shows the images captured for the reduction of the line to 3.53.5 cm which corresponds to a reduction of the 25%25% of the width. The captured images of the line shown in Figure 15 corresponds to the overpass of the vehicle over a line with a width reduction equal to the 50%50% which is a width of 2.52.5 cm. Furthermore, in both cases it was also possible to see the high degradation of the line that makes this line almost not visible for human eyes but was successfully detected by the system.

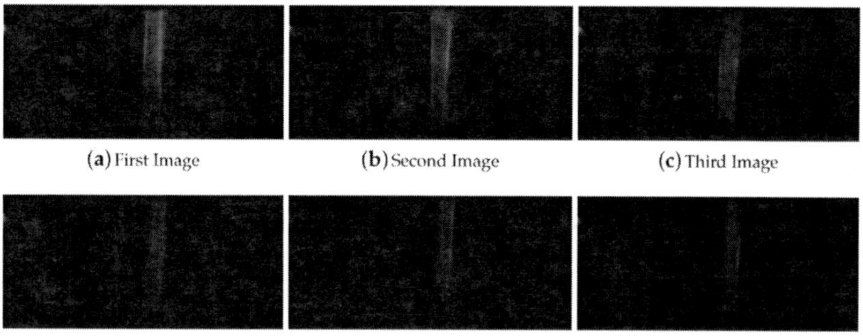

(a) First Image (b) Second Image (c) Third Image

(d) Fourth Image (e) Fifth Image (f) Sixth Image

Figure 14. Sequence of images for the reduction of the line width to 3.53.5 cm.

The presented line width reduction experiments were included in the circuit for the previously mentioned test of 13 laps shown in Figure 13. For each lap of this test the each line width reduction was overpasses one time.

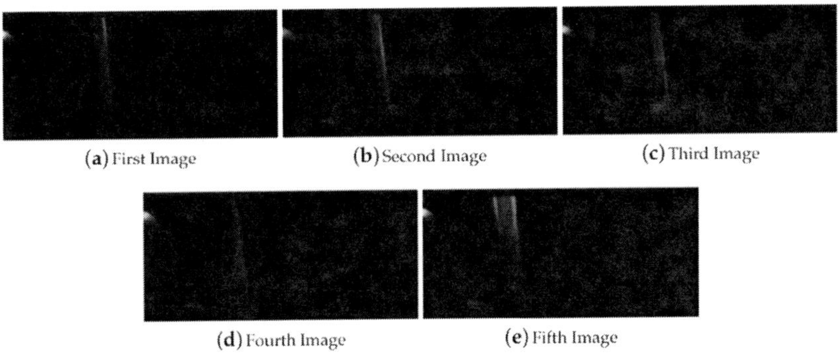

Figure 15. Sequence of images for the reduction of the line width to 2.52.5 cm.

6.1.1.3 Multiple Lines Simultaneously

In a potential real application of this system it could happen that the computer vision algorithm have to deal with the detection of more than one line simultaneously. That situation could be generated by the presence of previous painted lines, crossroads, or branch roads were two or more lines become one or the main road is divided in two or more lines.

To avoid to follow other detected lines which are not the current one a voting system between all the detected lines was implemented, as it was previously mentioned in Section 4.1. This system used a temporal information obtained related to the position inside the image of the detected line in the past frames. This system helps to reject false positives and other lines detected. The quality of all the detected lines is also taken into account to identified potential old painted lines. Figure 16show a sequence of images in where two lines are detected and one of them is rejected as false positive.

Furthermore, the computer vision algorithm rejects all the lines which have an inclination of more than 45° with respect to the vertical of the image frame. This assumption was taken to avoid problems with the presence of other lines in perpendicular direction that could be found in a crossroad situation, when the system overpass other tracks.

Another situation that was studied in this work is the presence of branch or fork roads. Train systems use this kind of roads to modify the current trajectory of the system to a new track or simply to send it to a garage or specific place for storage or repair the system. Based on this idea we assume that two different situations could happen, one, when the main guided line is divided in two or more, and the other, when the system overpass the junction of two or more lines which become one. Both situations are handle in the same way. An extra information was codified in the visual marks related to it. This information shows if there is or not a fork in this road, and if so where is it, and which of the line has to be track. The present implementation of the system can handle the division of the guided line in three new lines. The value of the variable which manage this situation could be *left*, *center*, *right*, and the system will keep tracking the line on the desired location. This variable could be also modified in real time. Figure 17 shows an example of the second situation in where a fork is overpassed and two lines become one.

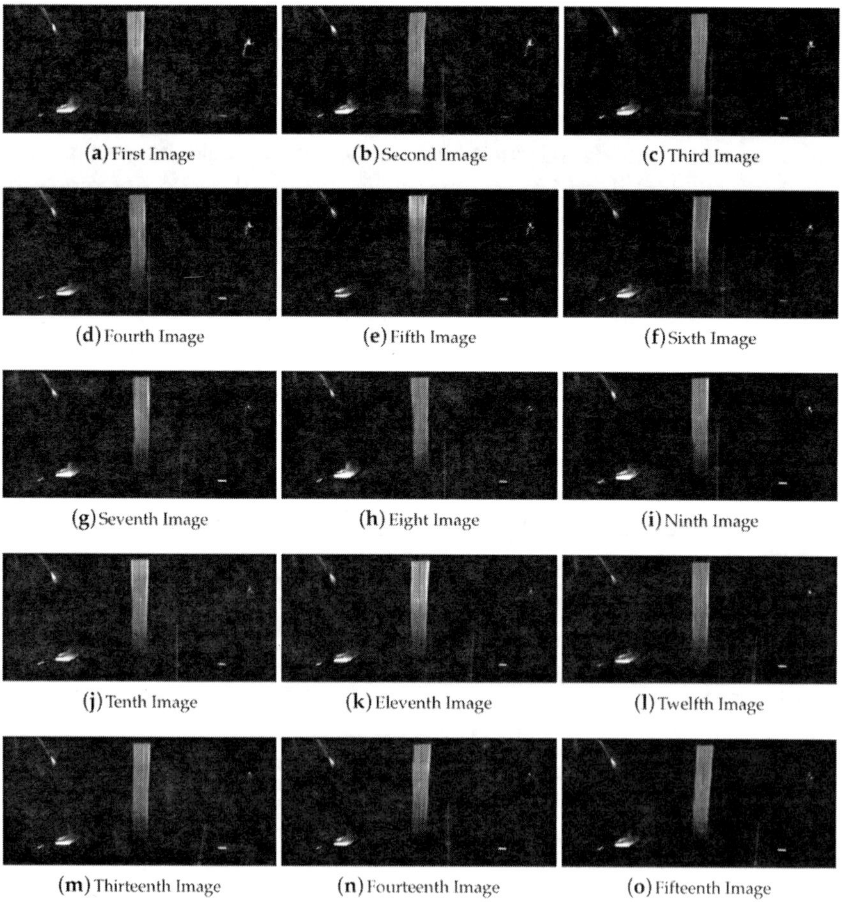

(a) First Image **(b)** Second Image **(c)** Third Image
(d) Fourth Image **(e)** Fifth Image **(f)** Sixth Image
(g) Seventh Image **(h)** Eight Image **(i)** Ninth Image
(j) Tenth Image **(k)** Eleventh Image **(l)** Twelfth Image
(m) Thirteenth Image **(n)** Fourteenth Image **(o)** Fifteenth Image

Figure 16. Sequence of images captured with a rejected false positive line.

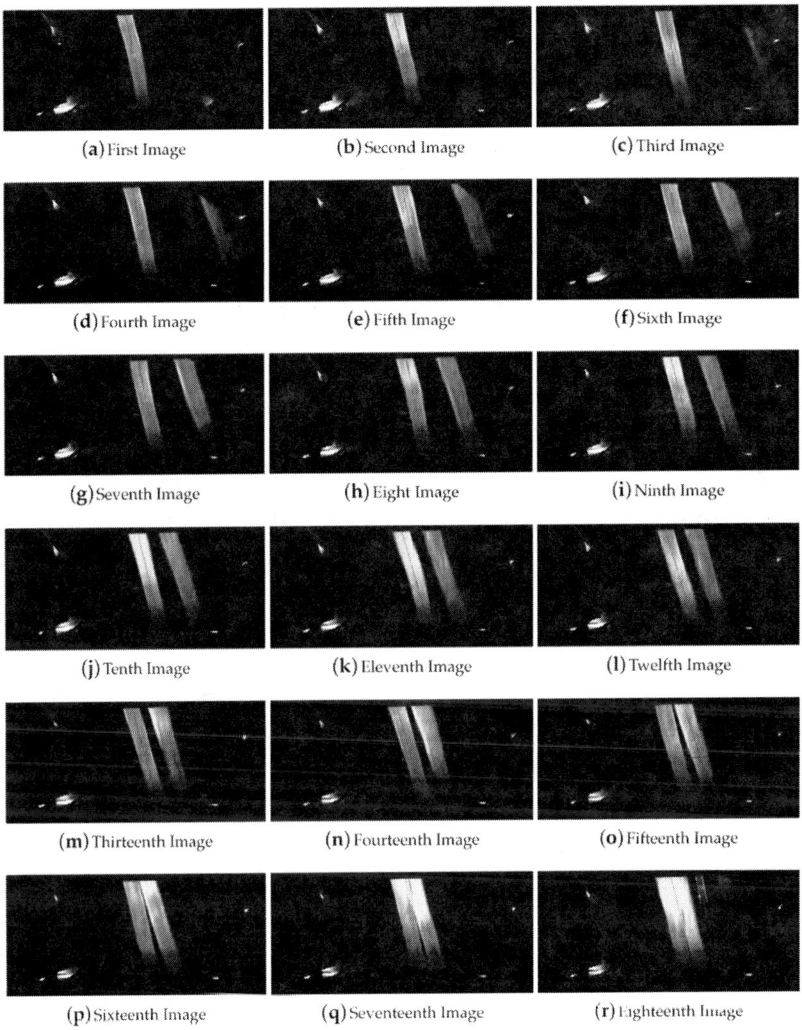

Figure 17. Sequence of images captured overpassing a fork or branch road.

6.1.2. Visual Marks Detection Robustness

6.1.2.1 Visual Marks Detection under Partial Occlusion

The robustness of the system in relation to the visual marks detection and its potential occlusions was also tested in this work. The defined size of the visual marks is 100 cm. A occlusion of the 50%50% of the visual marks was tested. The occluded visual mark was integrated in the same experiments as the line occlusions which were presented previously in Figure 13. The occluded visual mark was overpassed at the speed of 15 km/h and it was successfully detected during the 13 laps. One of this detection is shown in Figure 18.

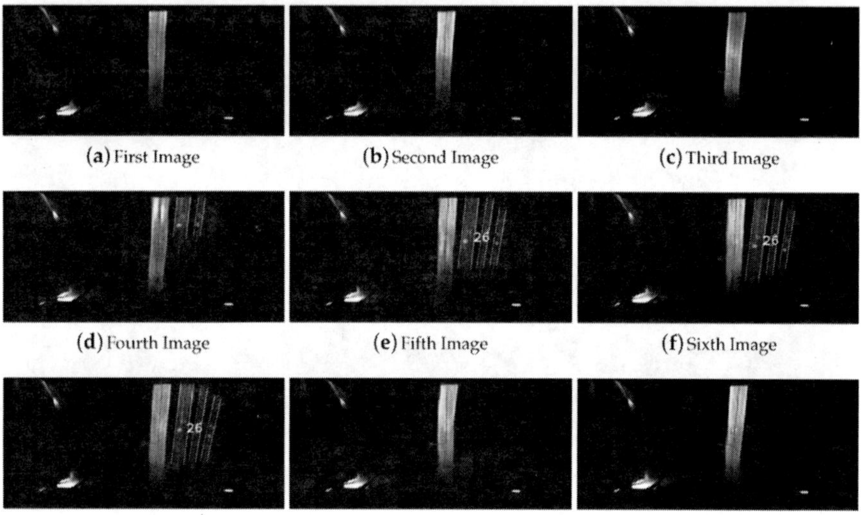

(a) First Image (b) Second Image (c) Third Image
(d) Fourth Image (e) Fifth Image (f) Sixth Image
(g) Seventh Image (h) Eighth Image (i) Ninth Image

Figure 18. Overpassing and detecting a half-size visual mark at the speed of 15 km/h.

6.1.2.2 False Detection: Wrong ID Identification

The robustness against false detection of visual marks is done based on two different voting process which was presented in Section 4. These two voting processes check the visual mark in different sections of each frame and based on the winning code detected in the previous frames. Figure 19 shows how the system initially identify the mark 25, then the visual mark 17 and after the mark id 25 again. The final detection and identification is not done until the third frame, when it is mentioned at the top left of the image by the message *Code: 25 (new!)*. The label *(new!)* shows that it is the first frame in which the visual mark was identified after the voting processes. This images sequence also shows the processed image on the bottom of each frame.

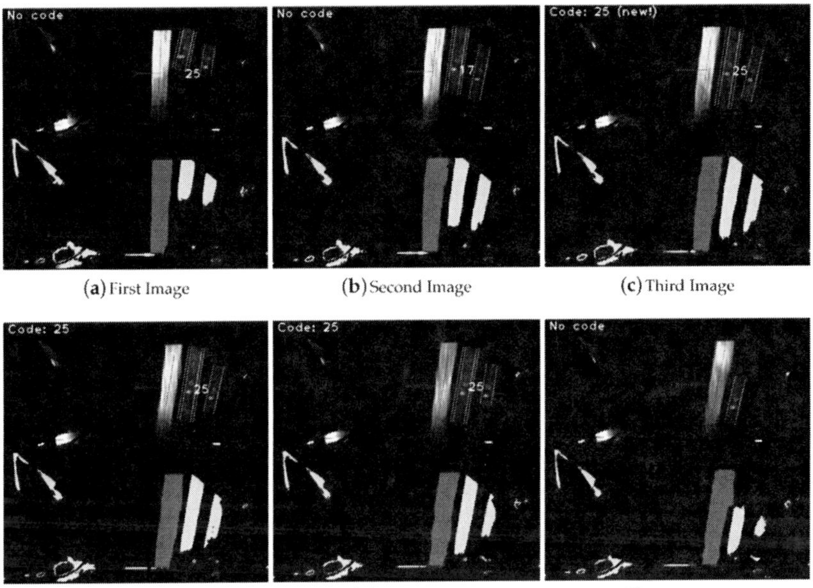

(a) First Image (b) Second Image (c) Third Image

(d) Fourth Image (e) Fifth Image (f) Sixth Image

Figure 19. Successful visual mark identification with a false detection of the visual mark 17.

6.2.2.3 False Detection: Wrong Detection

A false detection of a visual mark could be also caused by the detection of a code when there is no mark painted in the road. This could be caused by the presence of light reflections. These reflections are usually happen during the sun dawn and sunset, because the sun rays hit the road more horizontally and could enter into the metal structure where the camera is located. As is shown in Figure 20 the system could detect the sun reflections as visual marks but the codification system and the location definition of the marks, next to the guide line, make the computer vision system robust against this type of false detections too.

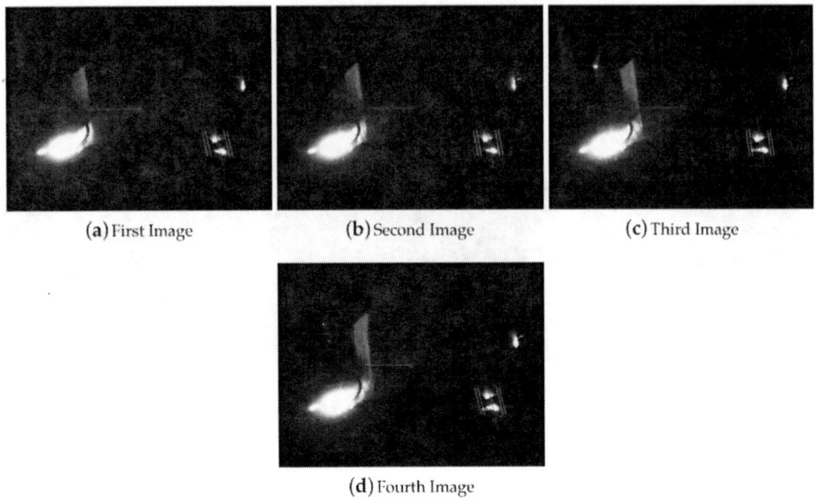

Figure 20. Visual mark wrong detection filtering.

6.1.3. Line and Visual Marks Detection at Different Speeds

There were no problems in the detection of the line at different speeds in the interval from 0 to 50 km/h. As set of experiments at different speeds are shown in Section 6.2.

6.1.4. Line and Visual Mark Degradation

The sunlight, rain and other weather factors can deteriorate the paint of the line and the visual marks, decreasing the quality of it and the pixels detected by the system. To detect the degradation process of the line and the marks, the system is evaluating in every captured frame the quality of it based on the number of pixels detected and the color values in predefined threshold intervals. These threshold intervals were set taking into account the acquired values on the experiments done in different weather condition and seasons during the set-up of the system.

6.2. Autonomous Driving

In this subsection it is presented the different experiments done to evaluate the behavior of the autonomous driving and speed assistance systems and the robustness of them.

6.2.1. System Stability

Initially, step response tests were done to check the robustness of the control approach against disturbances. These tests were defined changing the desired position of the line by 50 pixels. These step disturbance's tests were done in straights and curves at the different speeds of 10, 15, and 20 km/h. Table 1 shows the results of all the tests performed and the value of the Root Mean Squared Error (RMSE) for each test, which never were bigger that 8 cm.

Table 1. Evaluation of the system behavior against step disturbances of 50 pixels at different speeds and location inside the track.

Step Size (pixels)	Circuit Section	Speed (km/h)	RMSE (cm)
50	straight	10	7.5051
50	straight	10	6.8402
50	straight	15	7.8274
50	straight	15	6.7190
50	curve	10	6.9561
50	curve	10	5.1034
50	curve	15	5.5429
50	curve	15	6.5174
50	curve	15	6.2648
50	curve	20	6.9676

6.2.2. System Robustness

Once the system stability was tested, we checked the robustness of the system by different type of experiments. The results of the most relevant tests are shown in Table 2. This table is divided by the type of experiments into six different blocks. Following, it is mentioned separately the details of each type of experiments including the graphs of some of them. It has to be taken into account that the RMSE value is calculated from the beginning of the test until the end of it. Slightly different values of the RMSE for similar tests could be calculated based on the different acceleration/brake actions.

6.2.2.1 Final Control Adjustments

The first block shows the experiments done for the finest adjustment of the control system at different speeds between 10 and 20 km/h. Basically the part of the controller that was adjusted in these tests was the gain of the integral part. The gain adjustment was done in real time, testing one different value per each lap. A post-evaluation of the results and the error graphs were done to set-up the final value of the gain. The Figure 21 shows Experiment 2 mentioned in Table 2. In Figure 21b is shown the evolution of the error and slight differences for each lap.

6.2.2.2 Long Distance Robustness

Once the control system was completely set-up, the robustness of the system for long distances at low speed was tested. In this case the speed of the system was set from 10 to 16 km/h. The system was successfully tested in covering a distance of 7.37.3km, which is equivalent to 30 laps of the circuit at the speed of 15 km/h and 21 laps at different speed between 10 and 16 km/h. The system accomplished successfully both experiments with a RMSE error smaller than 4 cm. The experiment in which the vehicle was covering the distance of 7.67.6 km with the autonomous driving and assistance speed system working together is shown in Figure 22. In this Figure it is shown the evolution of the speed value, the evolution of the position of the steering wheel and the error for the experiment of 30 laps. In this case the RMSE of the full tests was reduced to 3.68743.6874 cm. Analyzing the plot shown in Figure 22b the localization of the vehicle inside the track can be intuited, being the vehicle in a straight line when the steering wheel is around 0 degrees. The curve of 11 m of radius is represented by the maximum turn of the steering wheel around 250 degrees, and the bigger curve with 20 m of radius is represented by the small turn around 150 degrees.

Table 2. RMSE evaluation of the different experiments done.

Test id	Number of Laps	Min Speed (km/h)	Top Speed (km/h)	RMSE (cm)	Detailed Information
1	9	10	15	4.3270	Small control adjustments
2	13	15	20	3.5256	Small control adjustments
3	21	10	16	3.4100	Long distance robustness with small speed variations
4	30	15	15	3.6874	Long distance robustness at constant speed
5	13	15	15	5.8098	Line/visual-marks occlusions at constant speed
6	6	14	42	10.7916	Line/visual-marks occlusions at different speeds
7	4	6	38	8.6494	Human speed control + testing speed limits of the circuit
8	4	14	42	9.1397	Human speed control + testing speed limits of the circuit
9	6	14	24	8.4839	Speed assistance + testing speed limits for the big curve
10	14	10	23	7.5486	Speed assistance + testing speed limits for the small curve
11	17	15	25	5.5535	Speed assistance + stop/start and diff. speeds
12	2	10	48	3.0313	Top speed + two stop/start in curves

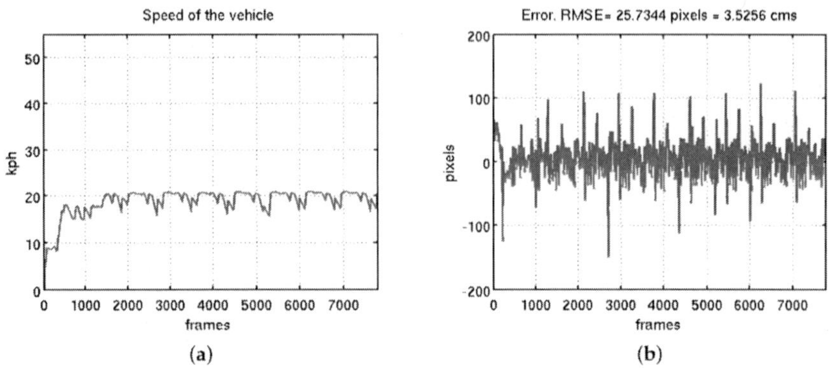

Figure 21. Finest adjustment of the steering wheel control system. (**a**) Vehicle speed evolution; (**b**) evolution of the error in pixels with a RMSE equal to 3.523.52 cm.

6.2.2.3 Line/visual Marks Occlusions

The third block of experiments are related to the evaluation of the system when there are line and visual marks occlusions. All these experiments were presented in the previous Section 6.1.

6.2.2.4 Speed Limits

The fourth block of experiments was designed to test the speed limits of the system under human speed control. In this case the speed assistance system was not active and a human was controlling the gas and brake pedals. Here is also checked the behavior of the system against faster accelerations and strong breaks. Table 2 shows two different experiments inside this block in which the speed was modified from 6 to 38 km/h and from 14 to 42 km/h. Figure 23 shows the evolution of the system in one of these tests. It is also checked in this test the behavior of the system when the vehicle starts non correctly aligned with the guided line. This could be checked in the first 20 s in Figure 23b,c by initial strong movements of the steering wheel and high error values.

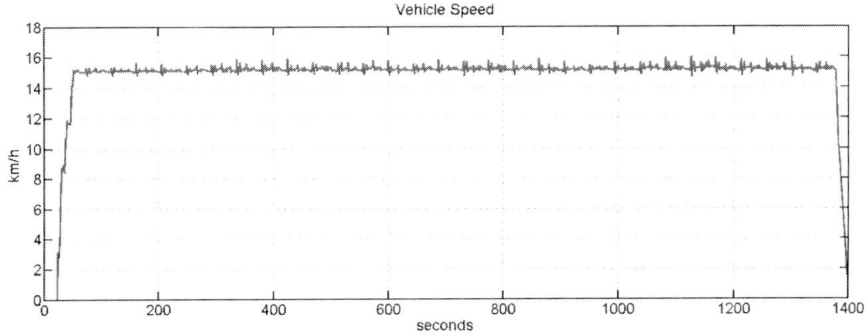

(a)

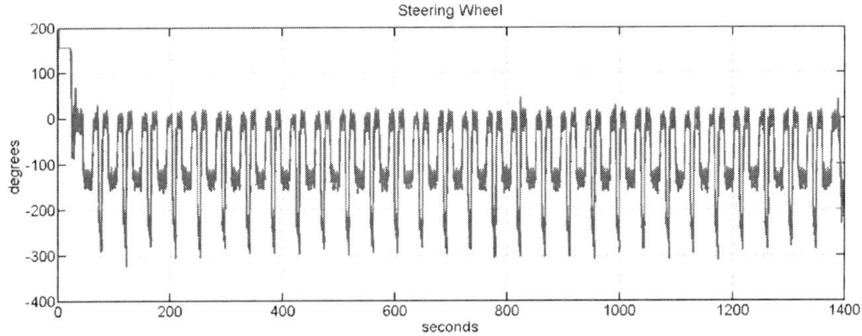

(b)

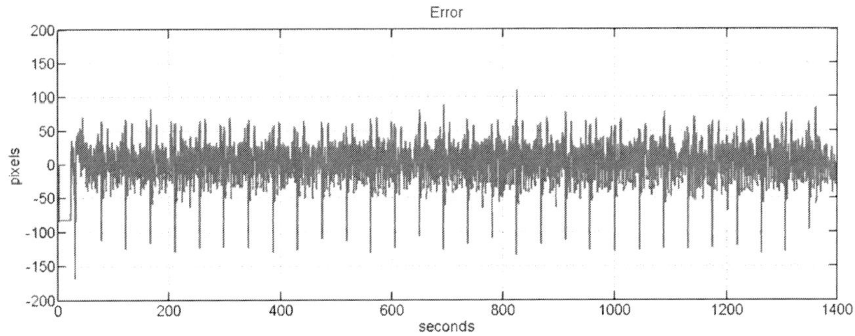

(c)

Figure 22. Evolution of the system to check the robustness of the system for a long distance with a reduced speed value. (**a**) Vehicle speed evolution; (**b**) evolution of the steering wheel; (**c**) error measured (in pixels), the value of RMSE for this test is 3.68743.6874cm.

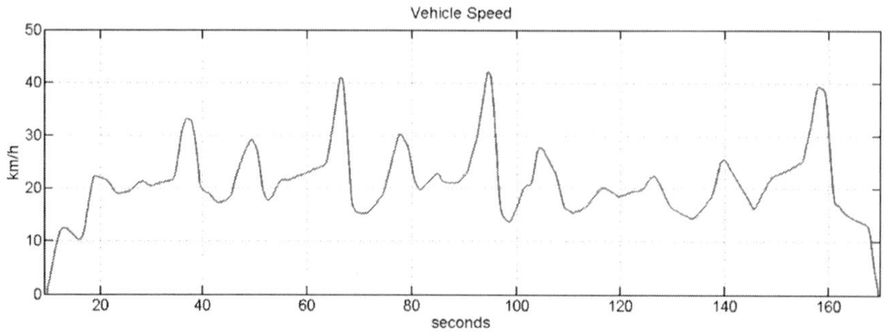

(a)

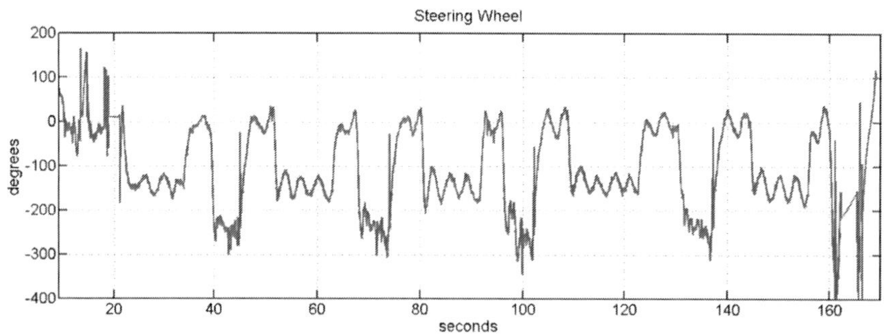

(b)

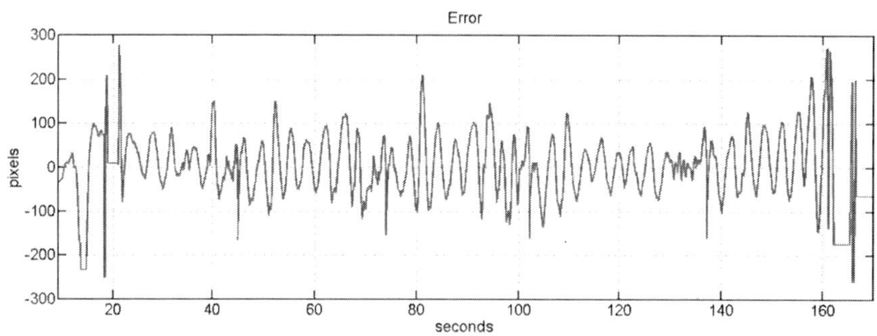

(c)

Figure 23. Evolution of the system under speed limits with human speed control experiment. (**a**) Vehicle speed evolution with a top speed of 42 km/h; (**b**) evolution of the steering wheel position; (**c**) evolution of the error measured in pixels, the value of RMSE for this test is equal to 9.13979.1397 cm.

6.2.2.5 Speed Assistance

In this block of experiments it was tested the performance of the speed assistance system. As is shown in the previous block of experiments, here it is also tested the speed limits of the system, but in this case we focus this action on the curves. The stop and restart of the vehicle based on visual mark detections was also tested here. These stops/restart actions try to emulate the stop of a public transport and were checked in curves and straights.

The behavior of the speed assistance system was also tested to response with an emergency stop of the vehicle. Here it is emulated the emergency situation by the detection of one specific visual mark. The system was configured to push the brake pedal until the end with the consequent strong deceleration. Figure 24 shows the behavior of the system on the emergency stop and also the evolution of the system when the speed is controlled by the speed assistance. In this test is also possible to see how the different sections are set with different speeds and how the speed assistance control it.Figure 25 shows another test of the emergency stop with the speed controller manually by the driver.

During the experiments presented on this and the previous blocks the comfort of the vehicle respect to the speed and the behavior of the lateral control system were tested at different speeds. The evaluation of these tests was done based in the feedback of some participants and the examination of the data obtained. Finally the speed limits of each section were set based on the extracted information.

6.2.2.6 Top Speed

The last block of tests focus on the robustness of the controller against big speed changes. In this test it is also checked the response of the system when the vehicle is traveling close to the maximum speed allowed in inner-city environment (50 km/h). In this case the speed was controlled by a human to applied a strong acceleration that permits to reach the maximum speed in the available straight access of the circuit. As is shown in Figure 26a, the autonomous vehicle reaches a top speed of 48 km/h. After this, a strong reduction of the speed was applied to enter in the biggest curve. Two stop/restart actions inside the smallest curve were also tested in this experiment. Figure 26c shows the error during these tests. The line was not lost in any of this strong stops and accelerations and the RMSE value along this experiment was equal to 3.03133.0313 cm. This value obtained represents just the 6%6% of the field of view of the perception system. The movements of the steering wheel commanded by the fuzzy controller is shown in Figure 26b.

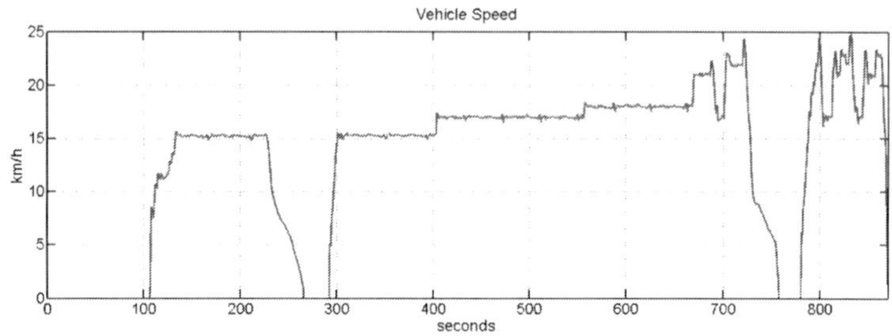

(a)

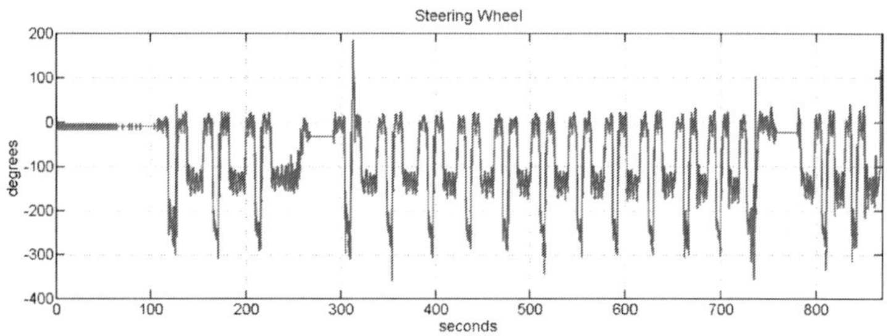

(b)

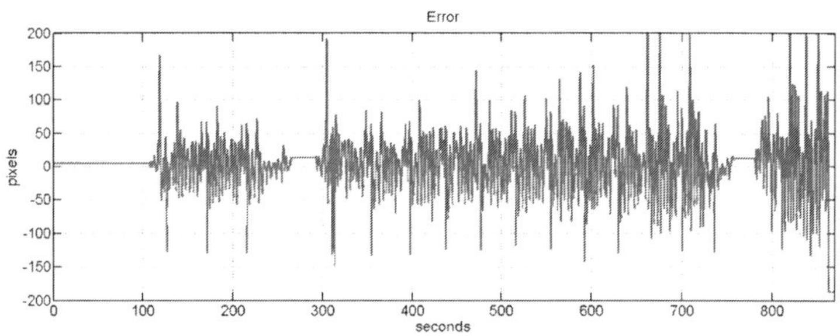

(c)

Figure 24. Evolution of the system with the speed assistance and the autonomous driving system working together and checking speed limits. It is also shown the response of the system to accomplish visual marks based stops. (**a**) Vehicle speed evolution; (**b**) evolution of the steering wheel; (**c**) evolution of the error measured in pixels, the value of RMSE for this test is equal to 5.55355.5535 cm.

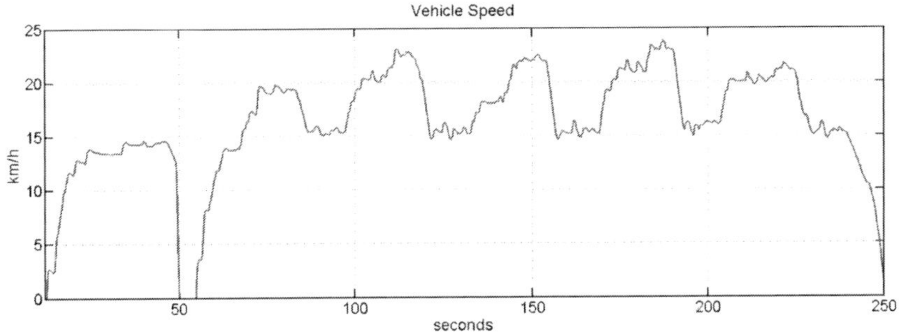

(a)

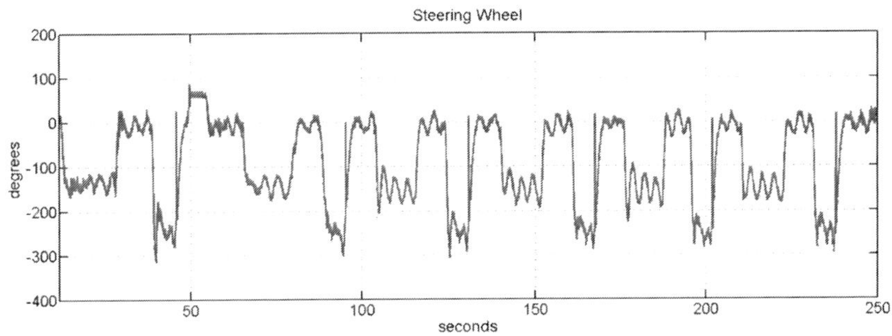

(b)

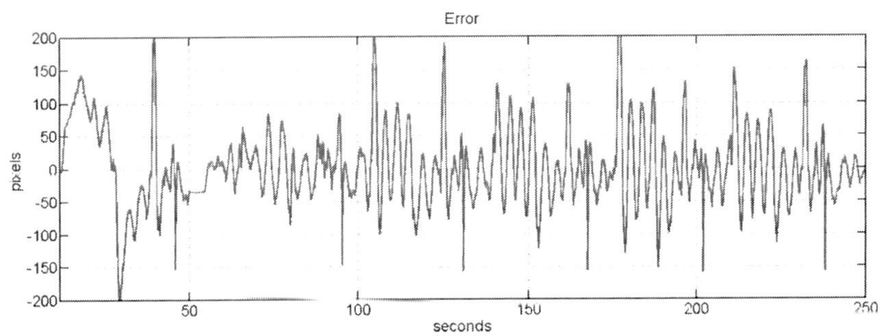

(c)

Figure 25. Evolution of the system with the speed assistance and the autonomous driving system working together. It is also shown the response of the system when a emergency stop is detected and done. (**a**) Vehicle speed evolution; (**b**) evolution of the steering wheel; (**c**) evolution of the error measured in pixels. The value of RMSE for this test is equal to 8.48398.4839 cm.

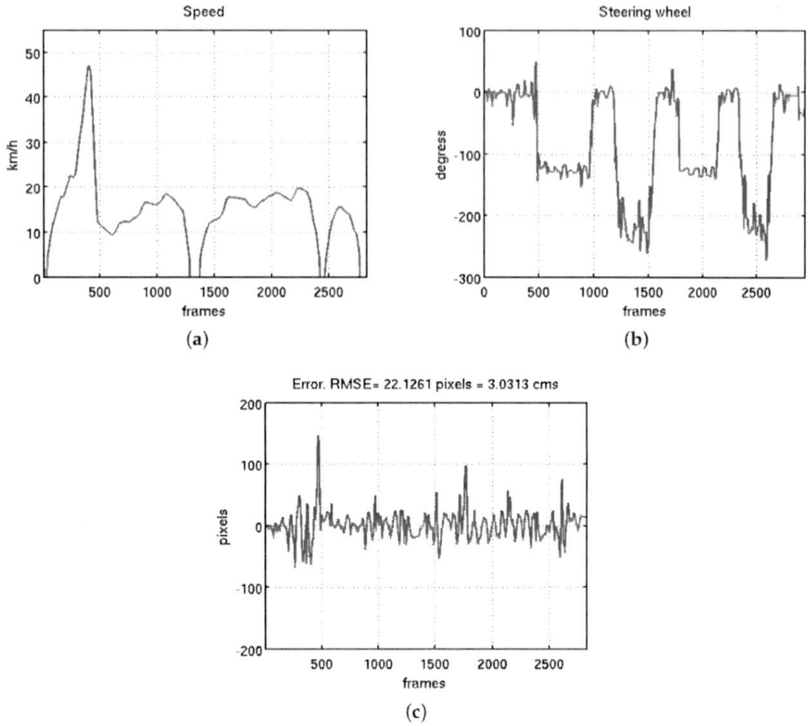

Figure 26. Evolution of the system under top city speed of 50 km/h and big speed changes. (**a**) Vehicle speed evolution with a top speed of 48 km/h; (**b**) evolution of the steering wheel; (**c**) evolution of the error measured in pixels, the value of RMSE for this test is equal to 3.03133.0313 cm.

The robustness of the system against line loss had been also tested and it is presented in details in [47]. In this previous work, no mark information was given to the control system with a maximum speed reduced. Successful results had been obtained following the same track without any information with a average speed of 15 km/h during more than 15 laps covering more than 4 km. Some tests of step commands in straight and curves at 10 and 15 km/h are shown too.

A video of the system working is available online [48].

7. CONCLUSIONS AND FUTURE WORK

The presented work shows a vision-based control system approach to command the steering of a passenger vehicle to follow a route. Furthermore, an assistance speed system and an approach to localize the vehicle inside the route based on a single vision sensor is also presented. The vision sensor was located inside a metal structure with a limited size of 30×5030×50cm. The definition of the path to follow was done by a line painted on the floor using ultraviolet sensible paint. The information measured by the computer vision system using the painted line was used as an input by a mixed control system that commands the steering of the vehicle. This control system is formed mainly by a PD-like fuzzy-logic controller, as well as a feedforward controller whose inputs are the position of the vehicle in the circuit and the curvature of the current section of the circuit. This extra information is also provided by the computer vision system by means of visual marks also painted on the road with a special UV sensible paint. Furthermore, these visual marks provide information related to the speed limit of the vehicle in each section of the route. The full system was successfully tested in a real passenger car inside a closed circuit traveling for more than 7 km without a driver and any interruption. Furthermore, different experiments were presented to test a maximum speed of 48 km/h, emergency stops and strong speed changes. The robustness of the computer vision algorithms against line and visual mark occlusions, false detection and the presence of more lines was also presented in a set of experiments. Based on the results of all of the experiments presented in this work, the presented approach is able to command a vehicle in inner-city environments as public transport without taking into account the obstacle avoidance. This is in fact a very important point in which the authors are focusing their effort. Another important part that could be improved is the dependence of the vision structure to avoid the illumination changes. The authors are also working in parallel in developing vision approaches that could be robust against illumination changes.

ACKNOWLEDGMENTS

The work reported in this paper is the product of several research stages at the Computer Vision Group Universidad Politécnica de Madrid in collaboration with INSIA-UPM and Siemens España S.A. This project is partially funded by the Centre for the Development of Industrial

Technology (CDTI). The authors would like to thank the company SIEMENS España S.A. that has made possible the research described in this paper through several contracts and the people at LABIE (INSIA-UPM) for their support.

AUTHOR CONTRIBUTIONS

Miguel A. Olivares-Mendez developed the control system approach, contributed to image processing development and designed and realized all of the experiments. Jose Luis Sanchez-Lopez contributed to the control approach developing the extra offset, collaborated in the development of the image processing algorithm and collaborate in the experiments. Felipe Jimenez was in charge of the vehicle automation and collaborated in the the experiments. Pascual Campoy was the project leader and guided the full study. Amin Sajadi-Alamdari and Holger Voos contributed to the full manuscript writing and the proofreading process.

CONFLICTS OF INTEREST

The authors declare no conflict of interest.

REFERENCES

1. Nath, C. Autonomous Road Vehicles—POST Note, PN 443. Houses of Parliament. Parliamentary Office of Science and Technology, 2013. Available online: http://researchbriefings.parliament.uk/ Research Briefing/Summary/POST-PN-443(accessed on 4 October 2015).

2. Olivares-Mendez, M.A.; Campoy, P.; Mellado-Bataller, I.; Mondragon, I.; Martinez, C.; Sanchez-Lopez, J. Autonomous Guided Car Using a Fuzzy Controller. In *Recent Advances in Robotics and Automation*; Sen Gupta, G., Bailey, D., Demidenko, S., Carnegie, D., Eds.; Springer Berlin Heidelberg: Berlin, Germany, 2013; Volume 480, pp. 37–55.

3. Chen, H. Research on laser guiding instrument for industrial carrier robots. In *Sports Technology and Engineering*; Advances in Sports

Technology and Engineering; Luo, Q., Ed.; CRC Press: Boca Raton, FL, USA, 2015; pp. 287–290.

4. Jung, K.; Kim, J.; Kim, J.; Jung, E.; Kim, S. Positioning accuracy improvement of laser navigation using UKF and FIS.*Robot. Auton. Syst.* **2014**, *62*, 1241–1247.

5. Kim, J.H.; Yang, W.; Jo, J.; Sincak, P.; Myung, H. *Robot Intelligence Technology and Applications 3*; Springer International Publishing: Cham, Switzerland, 2015; Volume 345.

6. Li, S.; Wei, W.; Wang, R. Study on Control Structure for the Automated Guided Vehicle Base on Visual Navigation. In Proceedings of the 2015 27th Chinese Control and Decision Conference (CCDC), Qingdao, China, 23–25 May 2015; pp. 2515–2518.

7. Bengler, K.; Dietmayer, K.; Farber, B.; Maurer, M.; Stiller, C.; Winner, H. Three Decades of Driver Assistance Systems: Review and Future Perspectives. *IEEE Intell. Transp. Syst. Mag.* **2014**, *6*, 6–22.

8. Bertozzi, M.; Broggi, A.; Conte, G.; Fascioli, A.; Fascioli, R. *Vision-based Automated Vehicle Guidance: The Experience of the ARGO Vehicle*; Elservier: AMsterdam, The Netherlands, 1998.

9. Broggi, A.; Bertozzi, M.; Fascioli, A. Architectural Issues on Vision-Based Automatic Vehicle Guidance: The Experience of the ARGO Project. *Real Time Imaging* **2000**, *6*, 313–324.

10. Behringer, R.; Muller, N. Autonomous road vehicle guidance from autobahnen to narrow curves. *IEEE Trans. Robot. Autom.* **1998**, *14*, 810–815.

11. Shladover, S. PATH at 20 History and Major Milestones. In Proceedings of the IEEE Intelligent Transportation Systems Conference (ITSC 2006), Toronto, ON, Canada, 17–20 September 2006.

12. Meyer, G.; Beiker, S. *Road Vehicle Automation*; Springer International Publishing: Cham, Switzerland, 2014.

13. DARPA Challenge, 2015. Available online: http://www.theroboticschallenge.org (accessed on 13 March 2015).

14. Thrun, S.; Montemerlo, M.; Dahlkamp, H.; Stavens, D.; Aron, A.; Diebel, J.; Fong, P.; Gale, J.; Halpenny, M.; Hoffmann, G.; *et al.* Winning the DARPA Grand Challenge. *J. Field Robot.* **2006**, *23*, 661–692.

15. Montemerlo, M.; Becker, J.; Bhat, S.; Dahlkamp, H.; Dolgov, D.; Ettinger, S.; Haehnel, D.; Hilden, T.; Hoffmann, G.; Huhnke, B.; *et al.* Junior: The Stanford Entry in the Urban Challenge. *J. Field Robot.* **2008**, *25*, 569–597.

16. Rauskolb, F.; Berger, K.; Lipski, C.; Magnor, M.; Cornelsen, K.; Effertz, J.; Form, T.; Graefe, F.; Ohl, S.; Schumacher, W.;*et al.* Caroline: An Autonomously Driving Vehicle for Urban Environments. In *The DARPA Urban Challenge*; Buehler, M., Iagnemma, K., Singh, S., Eds.; Springer Berlin Heidelberg: Berlin, Germany, 2009; Volume 56, pp. 441–508.

17. Google Inc., 2010. Available online: http://googleblog.blogspot.com.es/2010/10/what-were-driving-at.html (accessed on 21 November 2010).

18. Jadranka, D.; Müller, B.; Meyer, G. *European Roadmap Smart Systems for Automated Driving*; European Technology Platform on Smart Systems Integration (EPoSS): Berlin, Germany, 2015.

19. Dickmann, J.; Appenrodt, N.; Bloecher, H.L.; Brenk, C.; Hackbarth, T.; Hahn, M.; Klappstein, J.; Muntzinger, M.; Sailer, A. Radar contribution to highly automated driving. In Proceedings of the 2014 11th European Radar Conference (EuRAD), Rome, Italy, 8–10 October 2014; pp. 1715–1718.

20. SARTRE Project, 2012. Accesible online: http://www.sartre-project.eu/en/Sidor/default.aspx (accessed on 18 March 2012).

21. Broggi, A.; Medici, P.; Zani, P.; Coati, A.; Panciroli, M. Autonomous vehicles control in the VisLab Intercontinental Autonomous Challenge. *Annu. Rev. Control* **2012**, *36*, 161–171.

22. Broggi, A.; Buzzoni, M.; Debattisti, S.; Grisleri, P.; Laghi, M.; Medici, P.; Versari, P. Extensive Tests of Autonomous Driving Technologies. *IEEE Trans. Intell. Transp. Syst.* **2013**, *14*, 1403–1415.

23. Kohlhaas, R.; Schamm, T.; Lenk, D.; Zollner, J. Towards driving autonomously: Autonomous cruise control in urban environments. In Proceedings of the 2013 IEEE Intelligent Vehicles Symposium

Workshops (IV Workshops), Gold Coast, QLD, Australia, 23–26 June 2013; pp. 109–114.

24. Funke, J.; Theodosis, P.; Hindiyeh, R.; Stanek, G.; Kritatakirana, K.; Gerdes, C.; Langer, D.; Hernandez, M.; Muller-Bessler, B.; Huhnke, B. Up to the limits: Autonomous Audi TTS. In Proceedings of the 2012 IEEE Intelligent Vehicles Symposium (IV), Alcala de Henares, Spain, 3–7 June 2012; pp. 541–547.

25. Li, Q.; Chen, L.; Zhu, Q.; Li, M.; Zhang, Q.; Ge, S. Intersection detection and recognition for autonomous urban driving using a virtual cylindrical scanner. *IET Intell. Transp. Syst.* **2014**, *8*, 244–254.

26. Chen, L.; Li, Q.; Li, M.; Zhang, L.; Mao, Q. Design of a Multi-Sensor Cooperation Travel Environment Perception System for Autonomous Vehicle. *Sensors* **2012**, *12*, 12386–12404.

27. Vélez, G.; Otaegui, O.; Ortega, J.D.; Nieto, M.; Cortés, A. On creating vision-based advanced driver assistance systems.*IET Intell. Transp. Syst.* **2015**, *9*, 59–66.

28. Morales, S.; Klette, R. *A Third Eye for Performance Evaluation in Stereo Sequence Analysis*; Springer Berlin Heidelberg: Berlin, Germany, 2009; pp. 1078–1086.

29. Yebes, J.; Bergasa, L.; García-Garrido, M. Visual Object Recognition with 3D-Aware Features in KITTI Urban Scenes.*Sensors* **2015**, *15*, 9228–9250.

30. Rezaei, M.; Klette, R. Look at the Driver, Look at the Road: No Distraction! No Accident! In Proceedings of the 2014 IEEE Conference on Computer Vision and Pattern Recognition, Columbus, OH, USA, 23–28 June 2014; pp. 129–136.

31. Sivaraman, S.; Trivedi, M.M. Looking at Vehicles on the Road: A Survey of Vision-Based Vehicle Detection, Tracking, and Behavior Analysis. *IEEE Trans. Intell. Transp. Syst.* **2013**, *14*, 1773–1795.

32. Satzoda, R.K.; Lee, S.; Lu, F.; Trivedi, M.M. Snap-DAS: A vision-based driver assistance system on a Snapdragon^{TM}embedded platform. In Proceedings of the 2015 IEEE Intelligent Vehicles Symposium (IV), Seoul, Korea, 28 June–1 July 2015; pp. 660–665.

33. Ros, G.; Ramos, S.; Granados, M.; Bakhtiary, A.; Vazquez, D.; Lopez, A.M. Vision-Based Offline-Online Perception Paradigm for Autonomous Driving. In Proceedings of the 2015 IEEE Winter Conference on Applications of Computer Vision, Waikoloa, HI, USA, 5–9 January 2015; pp. 231–238.

34. Alvarez, J.; Gevers, T.; Diego, F.; Lopez, A. Road Geometry Classification by Adaptive Shape Models. *IEEE Trans. Intell. Transp. Syst.* **2013**, *14*, 459–468.

35. Wang, B.; Fremont, V.; Rodriguez, S.A. Color-based road detection and its evaluation on the KITTI road benchmark. In Proceedings of the 2014 IEEE Intelligent Vehicles Symposium Proceedings, Deadborn, MI, USA, 8–11 June 2014; pp. 31–36.

36. Jochem, T.; Pomerleau, D.; Thorpe, C. Vision-based neural network road and intersection detection and traversal. In Proceedings of the 1995 IEEE/RSJ International Conference on Intelligent Robots and Systems 95. 'Human Robot Interaction and Cooperative Robots', Pittsburgh, PA, USA, 5–9 August 1995; volume 3, pp. 344–349.

37. Gaikwad, V.; Lokhande, S. Lane Departure Identification for Advanced Driver Assistance. *IEEE Trans. Intell. Transp. Syst.* **2014**, *16*, 1–9.

38. Maddern, W.; Stewart, A.; McManus, C.; Upcroft, B.; Churchill, W.; Newman, P. Illumination Invariant Imaging: Applications in Robust Vision-based Localisation, Mapping and Classification for Autonomous Vehicles. In Proceedings of the IEEE International Conference on Robotics and Automation (ICRA), Visual Place Recognition in Changing Environments Workshop, Hong Kong, China, 31 May–7 June 2014.

39. Jimenez, F.; Naranjo, J.E.; Gomez, O. Autonomous Manoeuvring Systems for Collision Avoidance on Single Carriageway Roads. *Sensors* **2012**, *12*, 16498–16521.

40. Naranjo, J.; Jimenez, F.; Gomez, O.; Zato, J. Low Level Control Layer Definition for Autonomous Vehicles Based on Fuzzy Logic. *Intell. Autom. Soft Comput.* **2012**, *18*, 333–348.

41. Jimenez, F.; Naranjo, J.E.; Gomez, O. Dispositivo universal para el control automatico de la direccion de un vehiculo.*Dyna Ing. Ind.* **2014**, *89*, 398–404.

42. Hui, F.; Ming, Y.; Ruqing, Y. Ground Feature Point Matching Based Global Localization for Driverless Vehicles. *Robots***2010**, *32*, 55–60.

43. Mondragón, I.; Olivares-Méndez, M.; Campoy, P.; Martínez, C.; Mejias, L. Unmanned aerial vehicles UAVs attitude, height, motion estimation and control using visual systems. *Auton. Robots* **2010**, *29*, 17–34.

44. Olivares-Mendez, M.; Mejias, L.; IMellado-Bataller, I.; Campoy, P. See-and-avoid quadcopter using fuzzy control optimized by cross-entropy. In Proceedings of the 2012 IEEE International Conference on Fuzzy Systems (FUZZ), Brisbane, QLD, Australia, 10–15 June 2012.

45. Siciliano, B.; Khatib, O. *Springer Handbook of Robotics*; Springer: Berlin, Germany, 2008.

46. Sanchez-Lopez, J.L.; Campoy, P.; Olivares-Mendez, M.A.; Mellado-Bataller, I.; Galindo-Gallego, D. Adaptive Control System based on Linear Control Theory for the Path-Following Problem of a Car-Like Mobile Robot. In Proceedings of the IFAC Conference on Advances in PID Control PID'12, Brescia, Italy, 28–30 March 2012.

47. Olivares-Mendez, M.; Mellado-Bataller, I.; Campoy, P.; Mondragon, I.; Martinez, C. A Visual AGV-urban Car using Fuzzy Control. In Proceedings of the 2011 IEEE International Conference on Automation, Robotics and Applicatios (ICARA), Wellington, New Zealand, 6–8 December 2011.

48. Computer Vision Group - UPM (vision4uav) Website, 2016. Available online: http://vision4uav.eu/?q =researchline/partialAutonomousCar (accessed on 5 January 2016).

Chapter 6

Robot Guidance Using Machine Vision Techniques in Industrial Environments: A Comparative Review

Luis Pérez [1,*], Íñigo Rodríguez [1,†], Nuria Rodríguez [1,†], Rubén Usamentiaga [2] and Daniel F. García [2]

[1] Fundación PRODINTEC, Avda. Jardín Botánico 1345, 33203 Gijón (Asturias), Spain

[2] Department of Computer Science and Engineering, Universidad de Oviedo, Campus de Viesques, 33203 Gijón (Asturias), Spain

ABSTRACT

In the factory of the future, most of the operations will be done by autonomous robots that need visual feedback to move around the working space avoiding obstacles, to work collaboratively with humans, to identify and locate the working parts, to complete the information provided by other sensors to improve their positioning accuracy,*etc*. Different vision techniques, such as photogrammetry, stereo vision, structured light, time of flight and laser triangulation, among others, are widely used for inspection and quality control processes in the industry and now for robot guidance. Choosing which type of vision system to use is highly dependent on the parts that need to be located or measured. Thus, in this paper a comparative review of different machine vision techniques for robot guidance is presented. This work analyzes accuracy, range and weight of the sensors, safety, processing time and environmental influences.

Researchers and developers can take it as a background information for their future works.

KEYWORDS

machine vision; 3D sensors; perception for manipulation; robot guidance; robot pose; part localization

1. INTRODUCTION

Since the end of the 18th century with the first Industrial Revolution through the introduction of mechanical production facilities powered by water and steam, factories have experimented big changes in their production systems [1]. The second Industrial Revolution, in the start of the 20th Century, introduced mass production based on the division of labor powered by electrical energy [2]. The third Industrial Revolution of the start of 1970s introduced the use of electronics and information technologies for a further automatization of production [3]. Nowadays, we are involved in the fourth Industrial Revolution, commonly called "Industry 4.0", based on cyber-physical production systems (CPS) and embracing automation, data exchange and manufacturing technologies. These cyber-physical systems monitor the physical processes, make decentralized decisions and trigger actions, communicating and cooperating with each other and with humans in real time. This facilitates fundamental improvements to the industrial processes involved in manufacturing, engineering, material usage and supply chain and life cycle management [4].

Inside each revolution, several milestones have been achieved incrementing the innovation level. For instance, inside Industry 4.0, robot-based automatization has experimented its own revolutions. The first robotic revolution was the industrial automatization, the second one was the introduction of sensitive robots for safe automatization, the third one was the mobility with mobile manipulators, and the fourth and the last one is based on intelligent and perceptive robot systems [5].

The European Commission has set as objective for the Horizon 2020 Work Programme to achieve the leadership in industrial technologies (*i.e.*, Public-Private Partnership Factories of the Future or PPP FoF [6]). For this purpose, process automation and decreased accident rates are both important. Productivity and safety were limited by manual processes in the traditional industry; automatization and intelligent robots drive modern industry towards efficiency, resulting in a rapid increase in productivity, major material and energy savings and safer working conditions. Manufacturing demonstrates a huge potential to generate wealth and to create high-quality and highly skilled jobs [7]. In fact, several technology enablers have been identified within PPP FoF, such as advanced manufacturing processes and technologies, mechatronics for advanced manufacturing systems, including robotics, information and communication technologies (ICT), manufacturing strategies, knowledge-workers and modelling, simulation and forecasting methods and tools. Moreover, the European Commission has identified robotics as a key issue due to its importance for European economy: Europe is one of the world leading regions in industrial robotics with a share of more than 25% of supply and use. It is expected that robotics growth reaches 32 B$ by 2016 and a direct impact on job creation as forecasts determine that each industrial robot needs at least four people to run, maintain and service it [8]. Robotics directly improves society and better living conditions as it addresses global concerns such as climate change, sustainable transport, affordable renewable energy, food safety and security, and coping with an ageing population. Some examples of European projects in this area are TAPAS [9,10], VALERI [11,12] and SYMBIO-TIC [13].

Vision systems are widely used in industry, mainly for inspection and quality control processes [14]. Their use has been increased in applications related to improving the safety of workers in the industrial environment and for robot guidance [15]. Robots need machine vision to move around the working space avoiding obstacles, to work collaboratively with humans, to identify and locate the working parts, to improve their positioning accuracy, *etc*. Depending on the objective, the vision system can be scene-related or object-related [16]. In scene-related tasks the camera is usually mounted on a mobile robot and applied for mapping, localization and obstacle detection. In object-related tasks, the camera is usually attached to the end-effector of the robot manipulator (eye-in-hand configuration), so that new images can be acquired by changing the point of view of the camera.

Industrial robots are able to move to a position repeatedly with a small error of 0.1 mm, although their absolute accuracy can be several mm due to tolerances, eccentricities, elasticities, play, wear-out, load, temperature and insufficient knowledge of model parameters for the transformation from poses into robot axis angles [17,18,19]. In the automotive industry the accuracy requirement for operations such as spot welding will be of the order of 1 mm. The aerospace industry provides a challenging environment to apply robotics as its accuracy requirements are at least a factor of ten- to twenty-fold higher [20]. Conventional robots are not capable of achieving this accuracy. To improve the accuracy, optical calibration methods, such as laser tracker systems, photogrammetry or vision systems with multiple high resolution cameras, are used to detect the spatial position of the tool tip and to correct the robot motion. The combination of a measurement system with a robot is an optimal solution as it makes use of the robot ability for precise movement and overcomes the accuracy deficiencies. Moreover, working parts can be positioned slightly different from what the robot is expecting. In order to successfully navigate a mobile robot, obtaining detailed information on its immediate environment is the main concern. If the collected data is correct and sufficiently detailed, the creation of a 3D appearance model is possible and developers are given the opportunity to create accordingly sophisticated navigation software.

At this point, machine vision techniques and robotics become the main actors in the industrial scenarios. Thus, this work is a comparative review of different machine vision techniques and their applications for robot guidance, studying and analyzing the state of the art and the different approaches. The suitability of each vision technique depends on its final application, as requirements differ in terms of accuracy, range and weight of the sensors, safety for human workers, acquisition and processing time, environmental conditions, integration with other systems (mainly the robot), and budget. Main challenges are found in textureless surfaces for the correspondence problem, lighting conditions that may cause brightness, occlusions due to the camera point of view, undetermined moving objects, *etc.* Other comparatives can be found in the literature although they are oriented to commercial purposes, centered in one vision technique, or focused on software and algorithms. They are also analyzed in this work.

The rest of the paper is organized as follows: Section 2 resumes the fundamentals of vision-based 3D reconstruction,Section 3 reviews a wide range of robot guidance applications using machine vision, Section 4 performs a comparative analysis according to the application requirements and discusses advantages and drawbacks, and finally main conclusions are found in Section 5.

2. FUNDAMENTALS OF 3D RECONSTRUCTION

Three-dimensional perception is one of the key technologies for robots. A 3D view of the surroundings of the robot is crucial for accomplishing navigation and manipulation tasks in a fully autonomous way in incompletely known environments. Moreover, tele-operation of robots requires a visualization of the environment in a human-readable way, which is important for an intuitive user interface. Thus, vision systems for robot guidance generally need to obtain 3D information.

Given a point in the scene, its corresponding point in the image can be obtained by mathematical models [21]. This is the direct problem. As it determines a set of parameters that describe the mapping between 3D points in the world coordinate system and the 2D image coordinates, this process is also known as camera calibration. The perspective projection of the world coordinates onto the image is generally modeled using the pinhole camera model [22]. Figure 1 shows a graphical representation. Using this model, the image of a 3D point, *P*, is formed by an optical ray passing through the optical center and intersecting the image plane. The result is the point *P'* in the image plane, which is located at a distance *f* (focal length) behind the optical center.

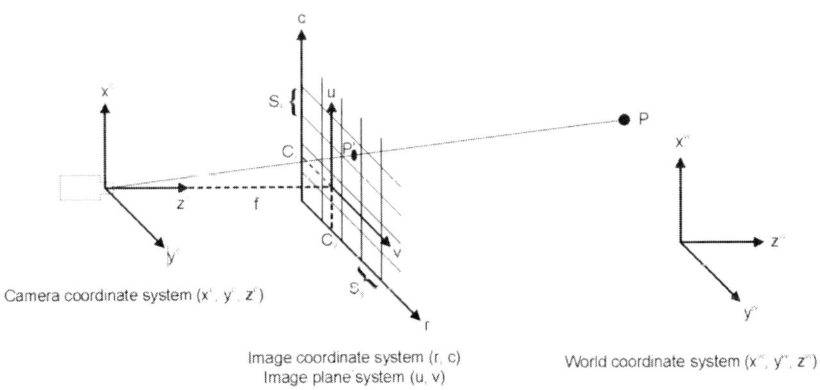

Figure 1. Pinhole camera model [23].

The first step to mathematically describe the projection of 3D points on the 2D image plane is the transformation from the world coordinate system *W* to the camera coordinate system *C*. This transformation is given by Equation (1). Using this equation, the camera coordinates of a point $P_c = (x_c, y_c, z_c)^T$ are calculated from its world coordinate $P_c = (x_c, y_c, z_c)^T$ using the rigid transformation $H_{w \to c}$:

$$\begin{pmatrix} p^c \\ 1 \end{pmatrix} = H_{w \to c} \begin{pmatrix} p^w \\ 1 \end{pmatrix} \tag{1}$$

The homogeneous transformation matrix $H_{w \to c}$ includes three translations *(t_x, t_y, tz)* and three rotations *(α, β, γ)*. These six parameters are called the extrinsic camera parameters and describe the rotation ($R_{w \to c}$) and translation ($t_{w \to c}$) from *W* to *C*. Thus, Equation (1) can also be expressed as Equation (2):

$$\begin{pmatrix} x^c \\ y^c \\ z^c \\ 1 \end{pmatrix} = \begin{pmatrix} R_{w \to c} & t_{w \to c} \\ 0 \quad 0 \quad 0 & 1 \end{pmatrix} \begin{pmatrix} x^w \\ y^w \\ z^w \\ 1 \end{pmatrix} \tag{2}$$

Based on the pinhole model, the projection of the point in the camera coordinate system *C* onto the image coordinate system is calculated using Equation (3):

$$\begin{pmatrix} u \\ v \end{pmatrix} = \frac{f}{z^c} \begin{pmatrix} x^c \\ y^c \end{pmatrix} \tag{3}$$

The pinhole model is only an ideal approximation of the real camera projection. Imaging devices introduce a certain amount of nonlinear distortion [24]. Thus, when high accuracy is required, lens distortion must be taken into account [25]. The final step is the transformation from the image plane coordinate system *(u, v)* to the image coordinate system *(r, c)*, which is the pixel coordinate system. This transformation is achieved using Equation (4), where S_x and S_y are scaling factors that represent the horizontal and vertical distances between the sensor elements on the CCD chip of the camera and the point $(C_x, C_y)^T$, which is the perpendicular projection of the optical center onto the image plane. Equation (4) reflects the calibration matrix:

$$\begin{pmatrix} r \\ c \end{pmatrix} = \begin{pmatrix} \frac{v}{S_y} + C_y \\ \frac{u}{S_x} + C_x \end{pmatrix} \qquad (4)$$

The projection of a point in the scene on the image can be mathematically calculated as shown in Figure 2a. However, given a point of the image it is not possible to obtain directly its original point in the space, as it is not a one-to-one relationship, but one-to-several, thus the inverse problem is ill-defined [26]. In algebraic terms, the projection of a 3D point on the image is not an injective application. Different points can be projected on the same pixel. What is really obtained by solving the inverse problem is a straight line formed by all points that are represented on the same pixel of the image. This is the projection line shown in Figure 2b.

Passive techniques, such as stereo vision or photogrammetry, which only require ambient lighting, solve the problem by looking for the same point in multiple images and computing the intersection of the projection lines. Others project a visible or infrared pattern onto the scene and estimate the depth information from the returning time (time of flight), the deformation of the pattern (light coding) or trigonometric calculations (laser triangulation and structured light). They are active vision. This difference in illumination method is important since the less well-defined features an object may have, the less accurate the system will be when passive vision is used. This is not the case with active vision systems, since a known pattern is used to illuminate the object. Nevertheless, using active vision can result in measurement inaccuracies, especially in the edges or in objects with varying surface finishes [27]. The main disadvantage of non-contact measurement systems is their high sensitivity to various external factors inherent to the measurement process or the optical characteristics of the object [28]. Passive vision techniques need multiple cameras for 3D reconstruction and active ones only use a single camera. However, as it will be shown, some techniques can be passive or active vision or use one or several cameras depending on the application. In addition, some of them are in fact evolutions or improvements of others. Thus, depending on the authors different classifications can be found over the literature. Table 1 shows one possible classification and points out that some technique can be included in more than one group.

Numerous sensor technologies are available, and each of them provide unique advantages for its use in specific environments or situations. In the following subsections, different 3D vision techniques are presented.

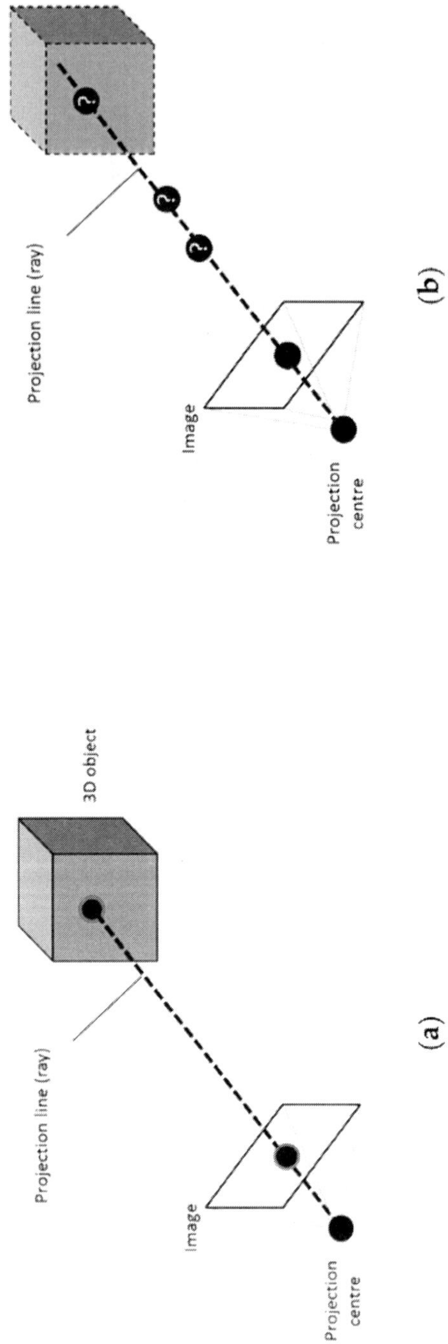

Figure 2. From 3D to 2D. (**a**) Direct problem; (**b**) Inverse problem.

Table 1. Vision techniques classification.

	Single Camera	Multiple Cameras
Passive vision	2D	Stereo vision Photogrammetry
Active vision	Time of flight Structured light Light coding Laser triangulation	Structured light Projected texture stereo vision

2.1. Stereo Vision and Photogrammetry

Literally photogrammetry consists of measuring real dimensions from a photo of an object. It is a 3D reconstruction technique based on conventional 2D images commonly used in architecture [29,30,31,32], topography [33,34], geology [35], archaeology [36], engineering [37,38], and manufacturing [39].

In stereo vision and photogrammetric techniques, the same point has to be found in other image to calculate the intersection of the projection lines and to obtain the 3D position (Figure 3). However, it is recommended that every point could be found at least in three images in order to ensure the detection and to improve the accuracy. The selected points must be the homologous and not any other ones in order to get the right 3D position.

Physical marks, such as stickers or laser points (Figure 4), are necessary over and around the object (the more, the better) and they must be of high contrast in order to ensure the detection. In Figure 5 two images of the same marked part have been taken and processed. The detected marks are printed in red. If the spatial position of the cameras and their calibration parameters are known, the marks can be paired using epipolar geometry and their projection lines and their intersections can be calculated to find the 3D position [40]. Notice that only those marked points are detected and used for the model (marker-based stereo vision).

In some cases, it is desirable to avoid these physical marks in order to save time and to automate the process. Feature tracking algorithms find, extract and match intrinsic characteristics of the objects between similar or consecutive images- avoiding physical marks, as can be seen in Figure 6 (markerless stereo vision).

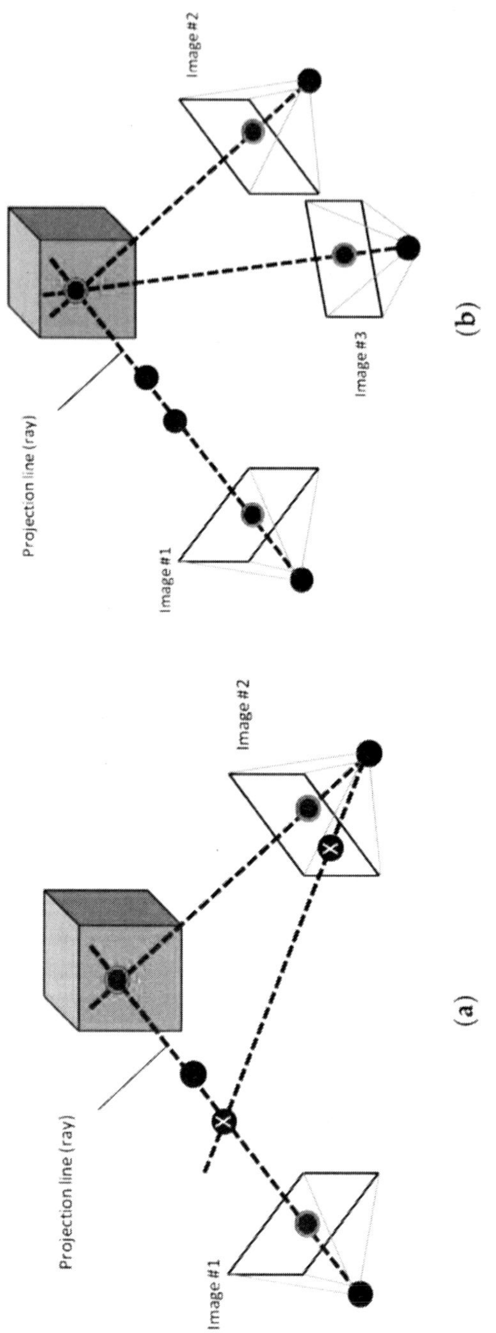

Figure 3. From 2D to 3D. (**a**) Homologous points; (**b**) Intersection of the projection lines.

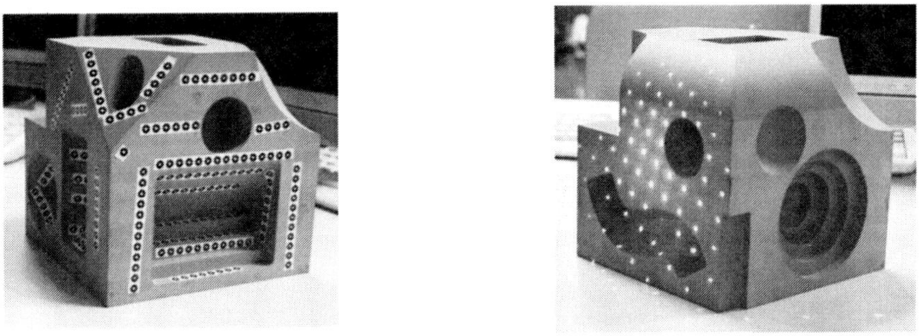

(a) (b)

Figure 4. Physical marks used in marker-based stereo vision. (**a**) Stickers; (**b**) Laser points.

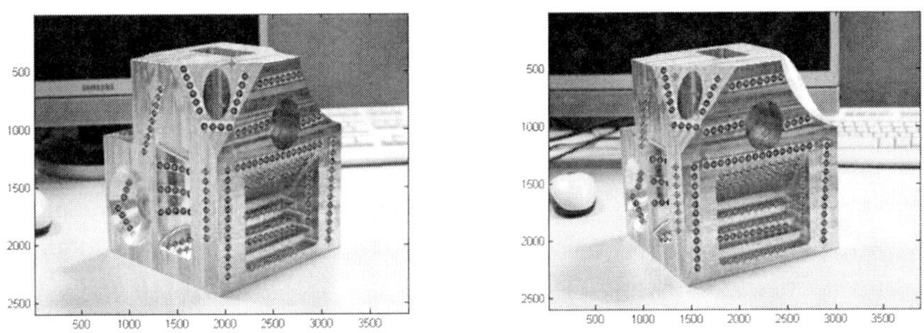

Figure 5. Detection of marks in several images.

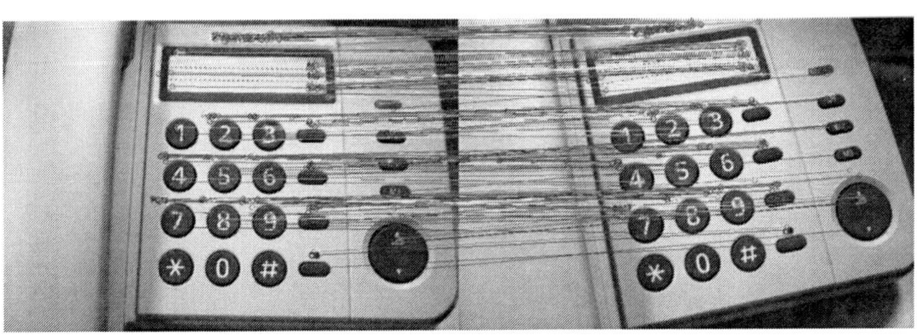

Figure 6. Feature tracking algorithms.

The extracted features depend on the problem or the type of application. In fact, a characteristic feature is a region of interest in the image that provides important information, such as an edge, a corner, bright or dark isolated points, *etc.* These detected points are useful for a subsequent search of correspondences between similar images. Some of the most popular algorithms for features detection, extraction and tracking are Canny [41], Harris [42], KLT [43], SIFT [44], SURF [45], and MSER [46].

Markerless stereo camera systems are widely used in many real applications including indoor and outdoor robotics. They provide accurate depth estimates on well-textured scenes, but often fail when the surface of the object is low-textured or textureless. In this case, it is necessary to project a known static high contrast light on it highlighting points, features, non-visible structures, *etc.* and creating an artificial texture. Then, the reflected light is captured using a stereo camera system and a matching algorithm associates the homologous points to obtain the 3D information [47]. *Ensenso* has developed several series of compact sensors based on this technique [48].

The projected texture is usually pulsed infrared light which is not affected by external light sources. It can take many forms including crosses, circles, squares, dot-matrices, multiple lines and random dot matrices. Finding the optimal texture, that is, the one which provides the best correspondence between features of the images, is a complicated problem, influenced by characteristics of the projector, the pattern, and the stereo cameras [49,50].

2.2. Time of Flight

Active vision techniques obtain the 3D information projecting a visible or infrared pattern on the object as shown inFigure 7. A time of flight (ToF) camera is a range camera that uses light pulses. The illumination is switched on for a very short time. The resulting light pulse is projected on the scene illuminating it and being reflected by the objects. The camera lens captures the reflected light onto the sensor plane. Depending on the distance, the incoming light experiences a delay which can be calculated as shown in Equation (5), where t_D is the delay, D is the distance to the object and c is the speed of light. The pulse width of the illumination determines the maximum range the camera can handle, thus the illumination unit is a critical part of the system. Only with some special LEDs or lasers it is possible to generate such short pulses.

$$t_D = 2 \cdot \frac{D}{c} \tag{5}$$

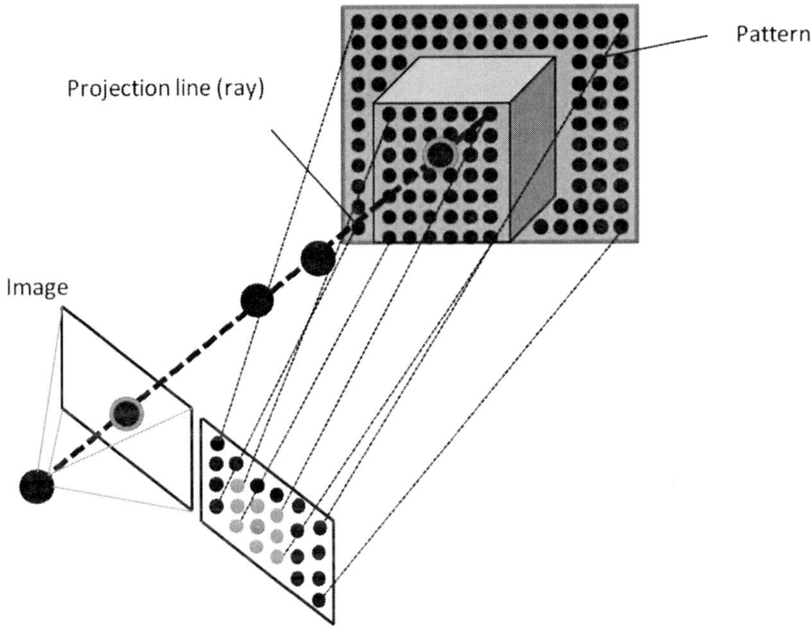

Figure 7. Projecting a pattern on the object.

One of the most common sensors has been developed by *MESA Imaging* [51]. Although ToF sensors open new possibilities and applications as they are quite compact and lightweight, and allow real-time distance acquisition, the quality of raw data is quite noisy and prone to several types of disturbances. They involve major specific challenges: (1) Low resolution compared with other techniques; (2) Systematic distance error; (3) Intensity-related distance error, as the distance is influenced by the incident light; (4) Depth inhomogeneity mainly at object boundaries; (5) Motion artifacts leading to erroneous distance values; (6) Multiple reflections; and (7) Other general aspects of active systems [52].

2.3. Structured Light

Structured light equipment is composed of a light source (the light projector) and one or two information receptors (the cameras). Among all the structured light techniques, there are two main groups [53]: time-multiplexing techniques, which are based on projecting a sequence of binary or grey-scaled patterns, and one-shot techniques, which project a unique pattern. The advantage of time-multiplexing techniques is that, as the number of patterns is not restricted, a large resolution,*i.e.*, number of correspondences, can be achieved. However, their main constraint is that the object, the projector and the camera must all remain static during the projection of the patterns. In one-shot techniques a moving camera or projector can be considered. In order to concentrate the codification scheme in a unique pattern, each encoded point or line is uniquely identified by a local neighbourhood around it [54].

Generally, the projected light is white light which is easily generated and is not dangerous for people unlike laser. This light is modified by grids to create lines or bands with lights and shadows like a zebra (Figure 8) which are recorded by the camera. Depth is obtained from the deviations using a technique similar to triangulation which consists on calculating the intersection between planes and lines [55].

Non-contact 3D digitizing scanners derived from structured light projection are increasingly more accurate, fastest and affordable [56]. Recently, new scanners have been launched based on blue light instead of white. As Figure 9 shows, they use a structured blue LED light module and a stereo scanner to generate the 3D point cloud. The LED module produces high contrast patterns allowing a high resolution scanning of the scenario. It is not affected by external light sources and it is safety for people. Sensors as the *HDI 109* and *HDI 120* from *LMI Technologies* can achieve an accuracy of 34 μm and 60 μm, respectively [57].

2.4. Light Coding

Recently a whole new type of sensor technology called light coding has become available for purchase at only a small fraction of the price of other 3D range finders. Light coding uses an entirely different approach where the light source is constantly turned on, greatly reducing the need for precision timing of the measurements. It can be considered an evolution of structured light.

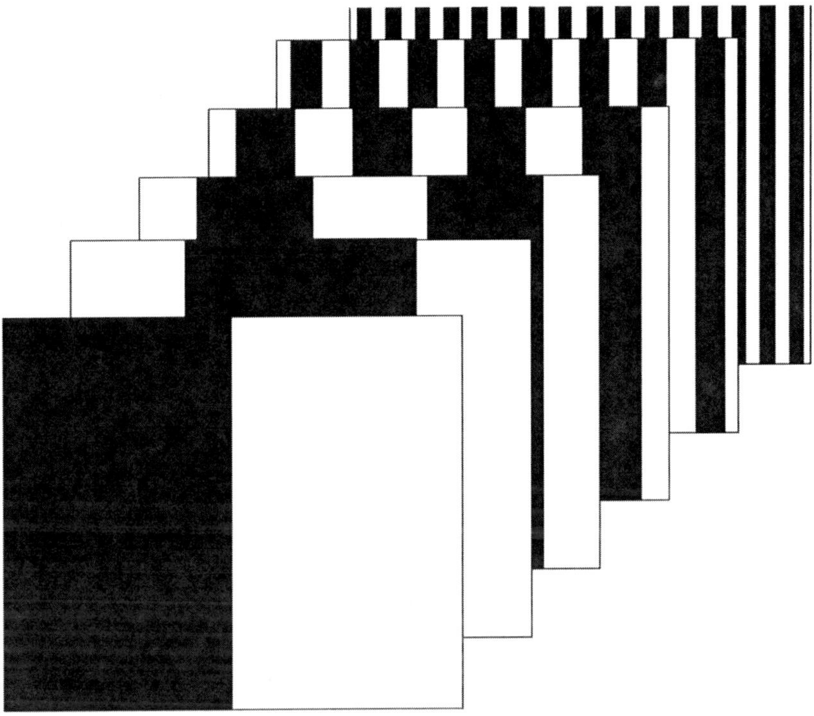

Figure 8. Structured light typical patterns.

A laser source emits invisible light (approximately at the infrared wavelength) which passes through a filter and is scattered into a semi-random but constant pattern of small dots which is shown in Figure 10. The reflected pattern is then detected by an infrared camera and analyzed. From knowledge on the emitted light pattern, lens distortion, and distance between emitter and receiver, the distance to each dot can be estimated measuring the deformations in shape and size of the projected points [58]. The technique has been developed by the company *PrimeSense* and its most commonly extended sensor is *Microsoft Kinect*.

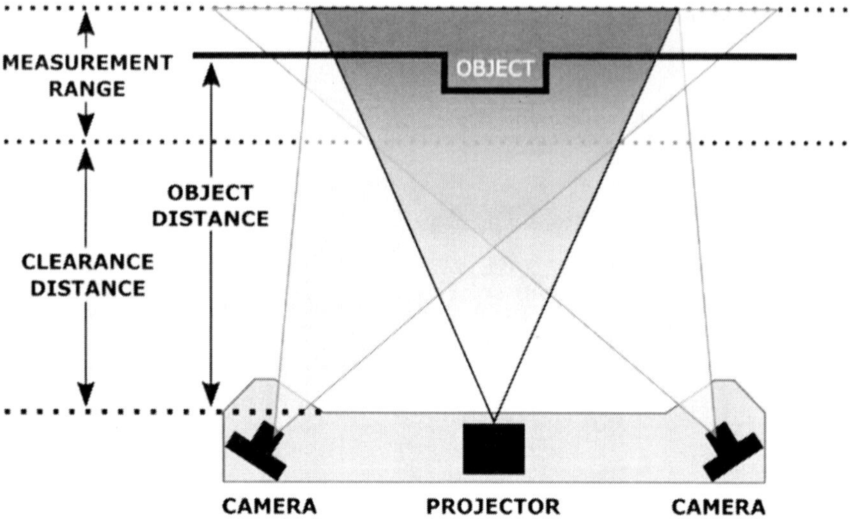

Figure 9. Blue LED sensor components.

Figure 10. Projected pattern in Light coding.

Light coding offers depth data at a significantly low cost, which is a great innovation not only for robotics. However, it has some limitations as these cameras do not provide a dense depth map. The delivered depth images contain holes corresponding to the zones where the sensor has problems, whether due to the material of the objects (reflection, transparency, light absorption, *etc.*) or their position (out of range, with occlusions, *etc.*). The depth map is only valid for objects that are in the range of 1–3 m in order to reduce the effect of noise and low resolution [59]. In addition to this, as it is based on an IR projector with an IR camera, and as the sun emits in the IR spectrum, sunlight negatively affects it.

2.5. Laser Triangulation

In laser triangulation, the point, the camera and the laser emitter form a triangle (Figure 11). The distance between the camera and the laser emitter is known, and because of the angle of the laser emitter corner is also known, the angle of the camera corner can be determined by looking at the location of the laser dot in the camera's field of view. These three pieces of information fully determine the shape and the size of the triangle and give the location of the laser dot corner of the triangle, which is in fact the 3D point. The accuracy depends on the resolution of the CCD sensor, the quality of the lenses, the point size (spot), the laser beam quality, the surface state of the piece and other optical factors [60].

Laser non-contact techniques have changed considerably in recent years. Although initially sensors gave isolated points, the technology has quickly spread to 2D sensors, measuring across a dotted line and collecting multiple points in a single frame. If the piece is moving under the sensor (or *vice versa*), the 3D model of the surface can be generated. This means that, unlike the previously presented techniques where a 3D point cloud can be obtained with a single capture, in this case it is necessary to move the piece or the sensor. In other words, it is necessary to scan the piece. As a drawback, depending on the laser power, laser sensors can be dangerous for people. They are not eye-safe.

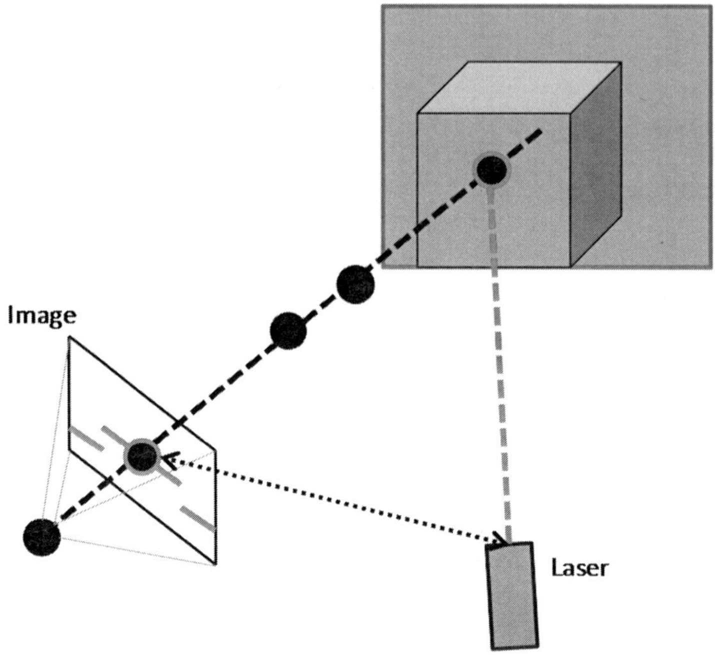

Figure 11. Laser triangulation.

3. ROBOT GUIDANCE IN INDUSTRIAL ENVIRONMENTS

Robot guidance using machine vision techniques is a challenging problem as it consists on providing eyes to a machine which is able to move with high repeatability but low accuracy [18] in complex industrial environments with other moving objects or even human workers. This modern technology opens up wholly new possibilities although it also creates new and fairly complex challenges in safety design [61]. Textureless surfaces, lighting conditions, occlusions, undetermined or moving objects, among others, are critical issues which the vision system has to deal with. As shown, 3D point cloud acquisition for robot pose estimation, robot guidance or any other purpose can be achieved by applying many different sensors and techniques. Which one is best suited for each particular application really depends on the needs and requirements [57]. The spatial coordinates of a large number of points can be obtained almost instantaneously or in a few seconds, but they require further treatment. Point clouds must be processed using specific algorithms to identify the geometric elements of the parts to be detected, measured or positioned. Subsequently filtering operations, structuring or interactive segmentation of the point clouds must be carried out [62]. The quality and the robustness of the final application are determined by both processes: the point cloud acquisition and the subsequent treatment.

In this section several approaches of robot guidance using different machine vision techniques are reviewed. As it will be shown, applications are oriented to scene-related tasks for environment reconstruction, including people detection, or to object-related tasks for robot pose estimation and object reconstruction for manipulation or inspection. Figure 12 provides the most common terms of robots that will be mentioned throughout the text.

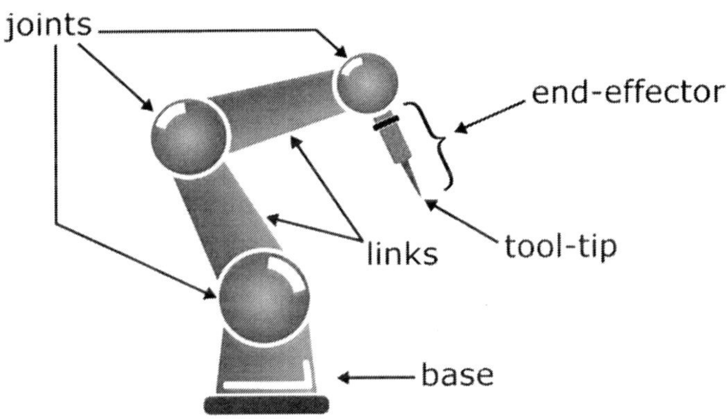

Figure 12. Robot terms.

3.1. Stereo Vision and Photogrammetry

Industrial photogrammetry covers different practical challenges in terms of specified accuracy, measurement speed, automation, process integration, cost-performance ratio, sensor integration and analysis. Main solutions are object-related in the fields of the measurement of discrete points, deformations and motions, 6 DOF (degrees of freedom) parameters, contours and surfaces [63]. Off-line photogrammetry systems can be found at automotive manufacturing (for car body deformation measurement, control of supplier parts, adjustment of tooling and rigs, *etc.*), the aerospace industry (for measurement and adjustment of mounting rigs, alignment between parts, *etc.*), wind energy systems (for deformation measurements and production control), and engineering and construction (for measurement of water dams, tanks, plant facilities, *etc.*). They offer high precise and accurate measurements (the absolute accuracy of length measurements is generally about 0.05 mm for a 2 m object [64]). On the other hand, on-line systems provide 3D information to control a connected process. Some examples include tactile probing (where a hand-held probing device is tracked in 3D space in order to provide the coordinates of the probing tip), robot calibration (where the robot tool center point is observed in order to determine its spatial trajectory), and sensor navigation (where a measurement device, such as a laser profile sensor, is tracked in 6 DOF in order to reconstruct the captured profiles). The accuracy of on-line systems is in the order of 0.2–0.5 mm over a range of 2 m [65], usually less than that of off-line systems. Nowadays, industrial photogrammetric systems are mostly used for off-line measurement of static 3D points in space. Moving from off-line to on-line systems is mainly a matter of speeding up image processing. Image acquisition, transfer, target identification and measurement usually take by far the largest part of the processing time.

There are several approaches in the literature to determine the position and orientation of a robot's end-effector with high accuracy during arbitrary robot motions based on combined and pure photogrammetric solutions. Laser tracking systems combine laser interferometry and photogrammetry [66,67]. As Figure 13 shows, the end-effector of the robot (the probe) is equipped with a number of LED reference targets suitable for camera imaging, as well as a retro-reflector suitable for laser tracking, all with calibrated local coordinates. The 6-DOF pose of the probe is measured by space resection through a camera that is integrated into the laser tracker. Distance information is provided by interferometric laser ranging while the camera measures angular information of the probe [68]. Laser trackers are also used to identify the geometric and dynamic parameters of the robot in order to improve the accuracy of the model, increasing thus the accuracy of the robot [69]. Qu [70] presents a laser tracker application to reduce the relative pose error of the robot of an aircraft assembly drilling process to less than 0.2 mm.

In pure photogrammetric solutions, there are three different approaches [18]: (1) Forward intersection, where two or more fixed cameras that are observing target points which are mounted on the end-effector (moving targets); (2) Resection, where alternatively one or more cameras can be mounted on the end-effector observing fixed targets; and (3) Bundle adjustment, which is a combination of both. At a first sight, resection arrangement may seem to be inferior compared to the forward intersection method, because a pose measurement for all possible robot poses requires targets to be placed around the entire workspace. However, most handling tasks require a robot to be accurate only in certain locations, thus the accuracy is not needed in the entire workspace.

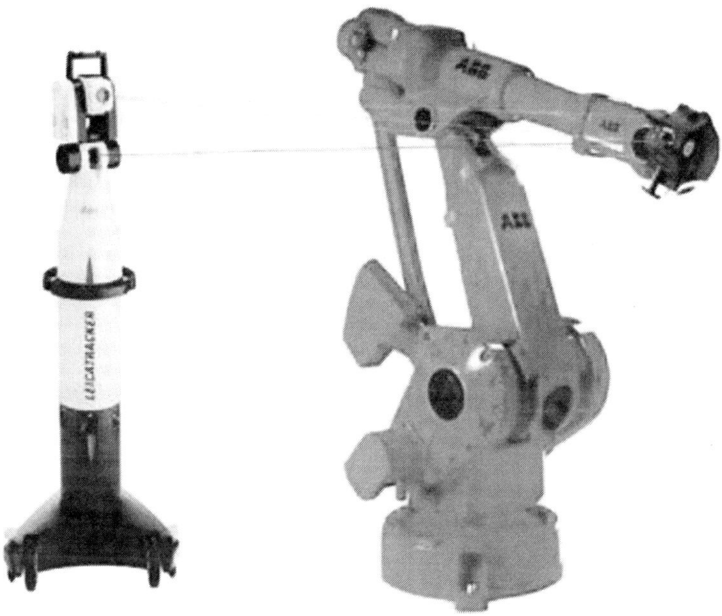

Figure 13. Laser tracker.

Main documented applications are for 6-DOF measurements, robot calibration, object tracking, and robot navigation. Several simultaneously operating solid-state cameras and photogrammetric hardware and software for robot guidance tasks are used in [71]. In this work two different applications were chosen to demonstrate the accuracy, flexibility and speed of the photogrammetric system: 3D object positioning is utilized in the measurement of car body points for accurate seam-sealing robot operation, and a robotized propeller grinding cell uses profile and thickness measuring data to control the grinding process. Hefele [18] shows an off-line photogrammetric robot calibration using a high resolution camera and targets mounted to the robot end-effector. The positioning accuracy of the robot is better than 3 mm in the experiments. First results towards an on-line photogrammetric robot tracking system are also presented in [18], reducing image processing and using intelligent cameras. The bundle adjustment indicates target coordinate RMS values of 0.06 mm in depth. Some improvements were added later in [72].

When accuracy requirements are moderate, [19] presents a digital photogrammetric system for industrial robot calibration tasks. Standard deviations of 0.05–0.25 mm in the three coordinate directions could be obtained over a robot work range of $1.7 \times 1.5 \times 1.0\ m^3$. In this sense, [20] describes the development of a photogrammetric 6-DOF measurement system mounted on the end-effector of an industrial robot. The functionality of the system has been demonstrated for drilling and assembly operations showing that the scheme is feasible and assesses the capability of such a system to operate within realistic tolerances. The photogrammetric system shown in Figure 14 is proposed in [73] to improve the pose accuracy of industrial robots. Experimental results show that the position error of the robot manipulator is less than 1 mm after being calibrated by the photogrammetric system. Amdal [74] gives a description of an on-line system designed for close range applications. The system has the ability to perform 3D point measurements with one single camera in combination with a measurement tool equipped with photogrammetric targets, which are precalibrated in the local tool coordinate system. Accuracy results obtained from simulation studies and measurement tests are reported. For a camera-to-object distance of 2.5 m, the accuracy was found to be 0.01 mm.

Consequently, photogrammetry is certainly able to determine robot pose accurately [75] although these measurements require special and expensive equipment and processing huge amounts of image data. Because of these reasons, nowadays photogrammetry is mainly limited for calibration, which is performed only once during the setup of the robot at the factory, instead of a continuous tracking of the pose. Recent and future developments are concentrated on higher dynamic applications, integration of systems into production chains, multi-sensor solutions and lower costs.

Projected texture stereo vision technique is mainly used in bin picking applications [76,77]. The task of picking random and unsorted objects from a container or a storage bin presents a number of different challenges. A fast and reliable identification of one or several objects is required in terms of shape, size, position and alignment (Figure 15). This information must be obtained and passed on to the robot controller, which is essential to the *ad hoc* generation of collision-free robot paths. This is the starting point for the use of robots in handling processes.

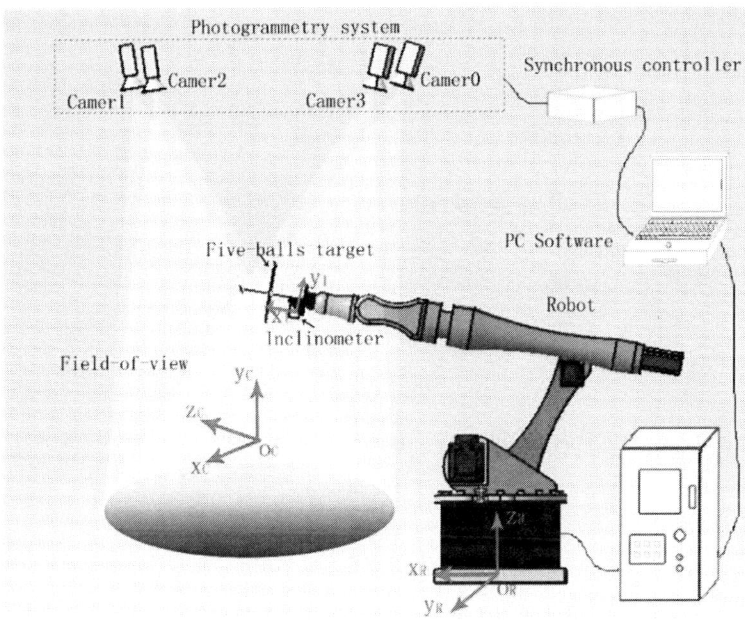

Figure 14. Multiple-sensor combination measuring system [73].

Sturm [78] presents an approach for detecting, tracking, and learning 3D articulation models for doors and drawers using projected texture stereo vision system. The robot can use the generative models learned for the articulated objects to estimate their mechanism type, their current configuration, and to predict their opening trajectory.

3.2. Time of Flight

ToF sensors have several advantages for the development of robotic applications as they are quite compact and lightweight, and allow real-time

3D acquisition with high frame rate. They are used in scene-related tasks, generally involving mobile robots and large displacements, and in object-related tasks, involving instead robotic arms or humanoid-like robots and small depths. However, they involve some challenges as the quality of raw data is quite noisy. To overcome this limitation, some authors apply calibration methods to rectify the depth images in order to obtain better results. Others complement ToF camera information with color cameras to create a 3D point cloud with real features, with grayscale cameras for redundant stereo or with laser scanners.

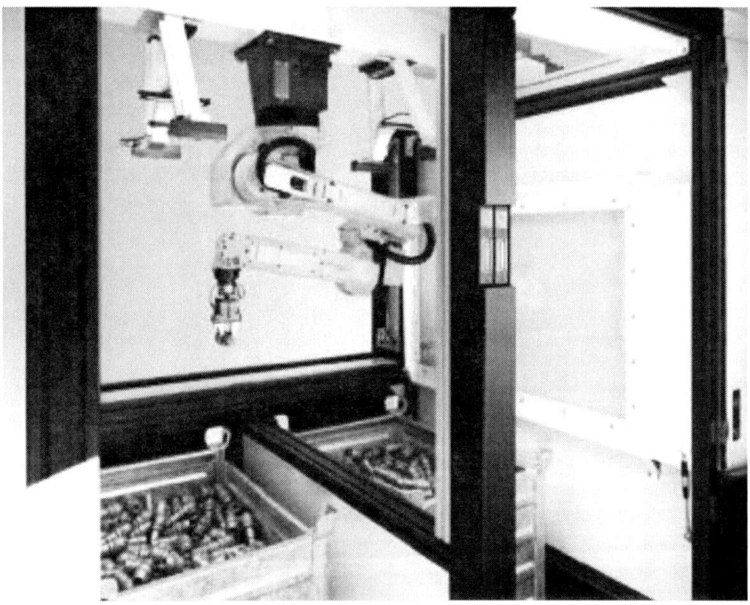

Figure 15. Bin picking [77].

In the field of scene-related task, ToF camera capabilities in terms of basic obstacle avoidance and local path-planning are evaluated in [79] and compared to the performance of a standard laser scanner. May [80] presents a new approach for on-line adaptation of different camera parameters to environment dynamics. These adaptations enable the usage reliably in real world (changing) environments and for different robotic specific tasks. In [81,82] it is proposed the use of surface normals to improve 3D maps for badly conditioned plane detection. Others, such as

[83,84], cope with ToF noisy point clouds using the Iterative Closest Point algorithm to find the relation between two point clouds. Arbeiter [85] performed an environment reconstruction for a mobile robot combining a ToF camera with two color camera, which is the input for a modified fast-SLAM algorithm. This algorithm is capable of rendering environment maps. Kuhnert and Netramai [86,87] combined a ToF sensor and a stereo system for environment reconstruction.

For object-related tasks, ToF cameras have also been successfully used for object and surface reconstruction, where the range of distances is smaller. Depending on the field of view of the camera, multiple 3D point clouds need to be acquired and combined. In fact, the most common setup usually includes a ToF camera mounted on the end-effector of a robotic arm to do the captures. Point cloud registration is more critical in object modeling than in scene modeling. Even if the hand-eye system is precisely calibrated, the displacement given by the robot is usually not enough and the transformation between different point clouds has to be calculated. Some examples of object modeling and object reconstruction can be found in [88,89,90,91]. For object manipulation, unknown and unsorted objects have to be identified or categorized in order to be grasped. Generally, it is not necessary to completely reconstruct the object. Some examples are described in [92,93,94].

Finally, for human-machine interaction, ToF does not require any special background and it is a non-invasive technique, contrary to the widely extended use of special gloves, artificial marks, special attached devices, *etc*. Thus, it is commonly used in human activity recognition than other vision techniques as [95] points out. This work reviews the state of the art in the field of ToF cameras, their advantages, their limitations, and their main applications for scene-related tasks, object-related tasks, and tasks involving humans.

3.3. Structured Light

Positioning a robot with stereo vision depends on features visible from several points of view. Structured light provides artificial visual features independent of the scene, easing considerably the correspondence problem. The main inconvenience for robot guidance is the size of the sensors as they include a projector, which makes them difficult to be attached to the end-effector of a robot. Figure 16 presents a solution where only one camera is attached to the end-effector (eye-in-hand) and a static projector is installed illuminating the working pieces [96].

Figure 16. Robot positioning using structured light [96].

The achieved accuracy is 3 mm, which is enough for this concrete application. This setup also solves a problem related to the breaking of the hot lamp filament of the project due to vibrations if it is moved around. On the other side, an eye-in-hand setup avoids occlusions while it can perceive more details during robot approaching to the scene. The selection of the adequate pattern is also the main focus of most authors. Pagès [54] proposed a coded structured light approach as a reliable and fast way to solve the correspondence problem in another eye-in-hand solution with the projector aside the robot manipulator. In this case, a coded light pattern is projected providing robust visual features independently of the object appearance for robot positioning. Experiments have demonstrated that positioning the robot with respect to planar object provides good results even in presence of occlusions. Results when using non-planar objects show that the camera motion is noisier, slower and less monotonic.

Le Moigne [97] describes some of the important operational considerations and image processing tasks required to utilize a non-scanning structured-light range sensor in path planning for robot mobility. Particular emphasis is placed on issues related to operating in ambient lighting, smoothing of range texture, grid pattern selection, albedo normalization, grid extraction, and coarse registration of images to the projected grid. The created range map can be converted to a topography map for planning short-range paths through the environment while avoiding obstacles.

The new approach of structured blue light is not very extended yet in industrial robotic applications, but it is already being used for part identification and localization in [11], where a mobile and collaborative robot has been developed for aerospace manufacturing industries. Once the robot has reached the working station, it takes the camera, acquires images to get a point cloud (Figure 17) and compares it with a CAD databank, in order to identify the part and its pose. The robot also corrects its own pose to start with the assigned task (apply sealant or do a quality inspection). In this case, the point cloud is highly accurate. However, as the sensor working distance is too short, the robot has to be very close to the part and the scanned area is small. For small parts this is not a problem, but for long ones it requires an accurate CAD matching algorithm in order to avoid deviations in the robot trajectory.

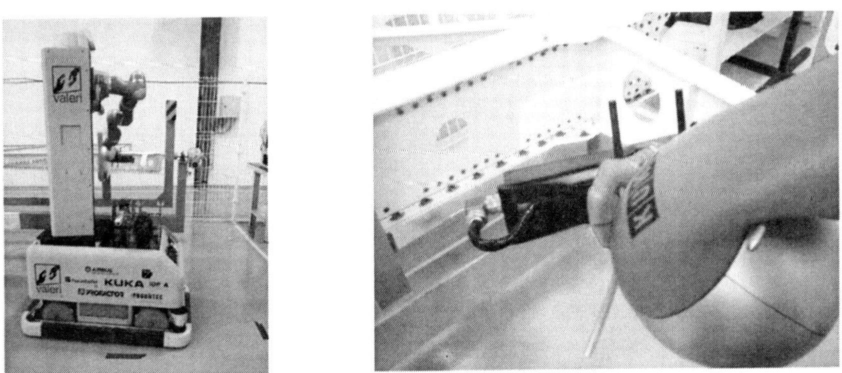

(a) (b)

Figure 17. Mobile robot using a blue light sensor for part localization. (**a**) Mobile robot; (**b**) Sensor operating.

3.4. Light Coding

Nowadays this technique is commonly used in videogames for people tracking. Besides, it is been introduced in more and more industrial applications including robotics although an extra effort is necessary to achieve accuracy for robot pose estimation, and safety requirements for workspace monitoring in human-robot collaboration. It offers visual and depth data at a significantly low cost and, while it is a great innovation for robotics, it has some limitations. Algorithms use depth information for object recognition, but as the sensor produces a noisy point cloud, it is required to improve such information. One possible option is the combination of light coding sensors with HD cameras to obtain high resolution dense point cloud which can be used for robot guidance or pose correction. Experimental results reported in [98] show that this approach can significantly enhance the resolution of the point cloud on both indoor and outdoor scenes. The point cloud is at least ten times denser than the initial one only with the light coding sensor.

To significantly improve the robustness of people detection on mobile robots, light coding cameras have been combined with thermal sensors and mounted on the top of a mobile platform in [99], since humans have a distinctive thermal profile compared to non-living objects. This approach also combines classifiers created using supervised learning. Experimental results of this work have shown that the false positive rate (exclusively achieved using only the light coding sensor) is drastically reduced. As a drawback, some phantom detections near heat sources, such as industrial machines or radiators, may appear. Light coding sensors are also combined with safety certificated 3D zone monitoring cameras in [61]. Wang [100] combined virtual 3D models of robots with the information from the sensor (images of operators) for monitoring and collision detection (Figure 18). 3D models, which are used to represent a structured shop-floor environment and linked to real motion sensors, are driven to mimic the behavior of the real environment. Light coding sensors add unstructured foreign objects, including mobile operators, which are not present in the 3D models.

One common problem of light coding sensors is that they do not provide a dense depth map. The delivered depth images contain holes corresponding to the zones where the sensor has problems, whether due to the material of the objects (reflection, transparency, light absorption) or their position (out of range, with occlusions).

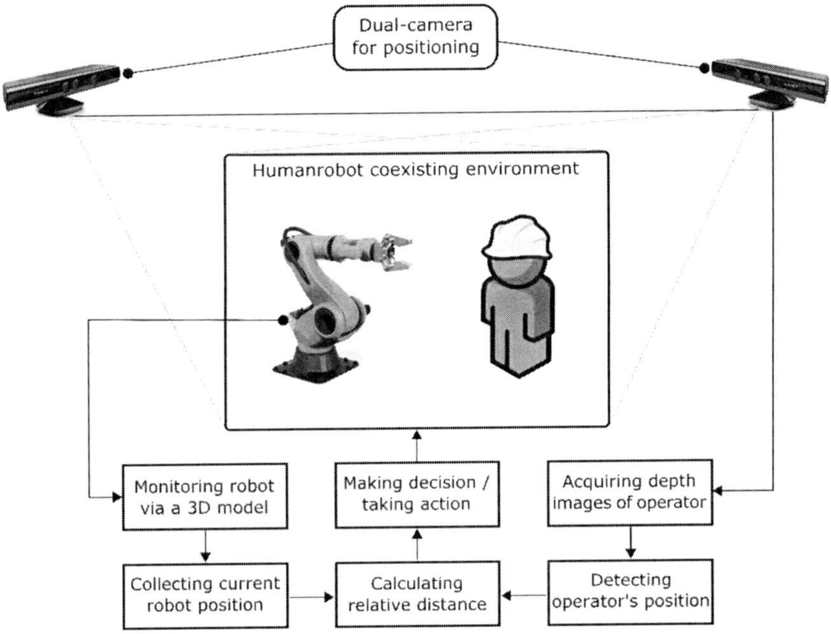

Figure 18. Combining 3D models of robots with information from sensors [100].

3.5. Laser Triangulation

This technique has been used in bin picking applications to pick up pieces, where it is necessary to recognize the piece and its pose. As it is necessary to scan the piece for 3D reconstruction, pieces are over a conveyor belt in [101] with a static camera and a static laser or camera and laser are integrated in robot tool for reconstruction and measurement as in [102]. Experimental evaluations of different line extraction algorithms on laser scans for mobile robots in indoor environments are presented in [103,104]. The comparison is carried out in terms of complexity, correctness and precision.

An implementation of a flexible, sensory-controlled robotic welding system is presented in [105]. Conventional, non-adaptive, robot welding systems can only be used when the workpieces are highly repeatable and well fixtured. A steerable cone of laser light and machine vision are used for sensing of the weld joint location and determining the detailed 3D weld joint surface geometry ahead of the welding torch. Robust vision-processing schemes for the detection and recognition of laser stripe features in noisy images are developed and implemented using a pipelined processing architecture. Approaches are proposed and implemented to incorporate the visually determined offsets in robot path planning and to control the welding process parameters. Another example of weld tracking is presented in [106]. This work offers a low-cost system that guarantees satisfactory tracking results even when the welding gap geometry varies strongly.

In [107] a sensor is created by coupling a camera and a laser stripe-. Positioning robotics tasks can be performed with good results, robustness and stability. Nevertheless, there are some constraints such as some restrictions in the laser stripe projection on to the scene. It is necessary to choose the most favorable location of the laser stripe during the calibration to achieve a robotics task under conditions of optimum stability. Pears [108] described a wide field of view range sensor for short range mobile robots maneuvers with an accuracy of 0.15% at l m, 1.3% at 2 m, and 3% at 2.5 m, and an average projected power of 0.9 mW, which is eye-safe. Generally, in robotics applications, it is necessary to take into account safety issues as human workers may be present. Depending on the laser power the sensor can result inappropriate for humans because of safety reasons.

Other works about robot navigation using laser scanners are described in [109,110,111]. In [109] the problems of self-localization of a mobile robot and map building in an *a priori* unknown indoor environment are addressed. In [110] a method for tracking the pose of a mobile robot using a laser scanner is presented. A new scheme for map building is proposed in [111]. This work describes localization techniques for a mobile robot equipped with a laser rangefinder using line segments as the basic element to build the map. According to the results, line segments provide considerable geometric information about the scene and can be used for accurate and fast localization.

4. DISCUSSION

Depending on the final goal of the application and the type of robot, different considerations and factors need to be taken into account in order to select the most adequate vision technique:

- *Accuracy* of point cloud alignment and *resolution*. They are mainly determined by hardware (sensor) and software (extraction, registration, segmentation, comparison, *etc.*), and in consistence to the size of the object and the application purpose.

- *Range of the sensor*. The working distance will be determined by the accessibility of robot, size of sensor and environment configurations

- *Light weight*. If the sensor is onboard or mounted in the end-effector, the robot has limited max load weight to ensure its full dynamics.

- *Safety issues*. The robot might work closely with human workers, thus sensors should avoid dangerous high-power laser to minimize any risk of accidents.

- *Processing time*. Processing time might be crucial to determine if a system is suitable for a certain application, especially regarding moving robots with safety constraints, *i.e.*, availability to detect and avoid collisions with humans or obstacles. Some techniques require that object and camera remain static for the capture, thus they are not applicable for moving scenarios.

- *Scanning environment*. Lighting conditions, vibrations, camera movements, *etc.* can disturb the quality of the 3D point cloud in some techniques. It is necessary to avoid these interferences.

- *Hardware and software integration* with other systems. The camera will be automatically controlled by the own robot central control unit or by an external source. *Ad hoc* developments are oriented towards the integration and, nowadays, most of current commercial vision systems are also prepared to be connected to a robot and controlled by external software using SDKs or libraries.

- *Budget*. Outside of technical issues, for a real implementation, budget should also be considered. A trade-off between cost and performance is necessary as most of the previous characteristics can be achieved or improved incrementing the invested amount of money.

A comparison of vision techniques is presented in Table 2 in terms of accuracy, range, weight, safety, processing time, and scanning environmental influences (Env. influences). Some quantitative information is provided according to the referenced specific application. Stereo vision, structured light, and laser triangulation can provide acceptable accuracy under certain conditions for most applications. Except structured white light, active vision techniques need to be closer to the object as they have a short working distance. Nowadays, there are light weight commercial sensors available to be mounted on a robot. Structured white light sensors are in general the biggest. All techniques are not dangerous for people, with the exception of some high power lasers that are not eye-safe. In terms of processing time, photogrammetry requires processing a large amount of images, and structured light techniques require that object and camera remain static during the acquisition process. Time of flight and structured blue LED light are not influenced by environmental lighting conditions.

Table 3 compares advantages and disadvantages of the reviewed techniques for robot guidance. Photogrammetry is mainly used in static applications because of its accuracy, but physical marks, such as stickers or laser points, are necessary (marker-based) and it is highly influenced by brightness and lights in industrial environments. Marks can be avoided using feature trackers (markerless), but the density of the point cloud would be low if surfaces are textureless. In fact, low-textured or textureless surfaces are also an inconvenience for stereo vision techniques with conventional 2D cameras and it is necessary to project a high contrast light creating an artificial texture to highlight points, features, *etc.* (projected texture stereo vision). Other 3D active vision techniques, such as light coding and time of flight, have low theoretical accuracy and are not valid for certain applications where the point cloud is compared with a CAD model for accurate part localization, because flat surfaces are represented with rather significant curvature. They can be used for part identification or for people and object tracking considering these accuracy issues. Laser techniques are commonly used in scanning applications where the capture is not a single snapshot, but can be dangerous for people as some laser classes (high power) are not eye-safe. Finally, structured light provides accuracy, although sometimes is influenced by ambient light and has problems to create the 3D model for surfaces of certain colors. Its main disadvantage is that most commercial sensors are quite big to be carried by a robot. Actually, research and development efforts are concentrated on miniaturizing sensors or on separating the projector and the sensor so that only one is onboard the robot. In this sense, the new evolution of this technique called structured blue LED light provides accuracy with a small sensor.

Table 2. Comparison of vision techniques in terms of accuracy, range, weight, safety, processing time, and environmental influences.

	Accuracy	Range	Weight	Safety	Processing Time	Environmental Influences
Stereo Vision and Photogrammetry	✓ (50 µm [64])	✓	✓	✓	✗ (image processing)	✗ (brightness)
Projected Texture Stereo Vision	✓ (0.1 mm [48])	✗ (0.25–3 m [48])	✓	✓	✗ (image processing)	✗ (brightness)
Time of Flight	✗ (10 mm [51])	✗ (0.8–8 m [16])	✓	✓	✓	✓
Structured White Light	✓ (0.127 mm [112])	✓	✗ (projector [96])	✓	✗ (remain static)	✗ (light, brightness)
Structured Blue LED Light	✓ (34 µm [57])	✗ (157–480 mm [57])	✓	✓	✗ (remain static)	✓
Light Coding	✗ (10 mm [58])	✗ (1–3 m [59])	✓	✓	✓	✗ (sun)
Laser Triangulation	✓	✓	✓	✗ (laser power)	✓	✗ (brightness)

Table 3. Advantages and disadvantages.

	Advantages	Disadvantages
Stereo Vision and Photogrammetry	Commonly used. Accuracy.	Influenced by environment. Physical marks are necessary. The density of the point cloud can be low. Object and camera must be static for the capture.
Projected Texture Stereo Vision	Physical marks are not required.	Influenced by environment. Object and camera must be static for the capture.
Time of Flight	Independent of ambient light. Not necessary that object and camera remain static for the capture.	Low theoretical accuracy.
Structured White Light	Accuracy.	Sometimes influenced by ambient light. Problems to create the 3D model for surfaces of certain colors. Expensive. Sensors can be quite large. Object and camera must be static for the capture.
Structured Blue LED Light	Accuracy. Small sensor.	Short working distance. Object and camera must be static for the capture. Expensive.
Light Coding	Inexpensive. Not necessary that object and camera remain static for the capture.	Low accuracy. Uncertified at industrial level.
Laser Triangulation	Commonly used. Inexpensive (depending on the laser, the accuracy).	Dangerous for people depending on laser power. Usually short working distance. Line scanner.

There are thousands of industrial applications but most of them are confidential for companies and are not documented or widely described in scientific papers, thus they could not be included in this survey. Table 4 summarizes the references of vision techniques for robotics reviewed in this work grouped by scene-related and object-related tasks. According to the survey, main stereo vision applications in robotics are in the field of object-related tasks for robot pose estimation and robot calibration as they may require marks, camera and object must remain static and an important amount of information needs to be processed. Time of flight, which does not require that object and camera remain static, is mainly used for environment and object reconstruction, navigation, obstacle avoidance and indeed people detection. In fact, the main application of light coding is people detection, although it is not certificated for industrial environments. Laser-based sensors are widely used especially for navigation, but also for object-related tasks.

Some other comparative reviews of machine vision techniques for robotic applications can be found in the literature. Wilson [27] reviews 3D vision techniques and solutions in a journal article with commercial purposes providing a list of applications and manufacturers. Sets of questions and advises are presented in [57] in order to help the reader to choose the right 3D vision technique for his/her project. Some of these questions are relative to the size and the surface of the target object, to the accuracy requirements, to the final destination of the obtained 3D data, or to the budget. This classification is quite similar to the one proposed by the authors of this work in Table 2.

Foix [95] focuses on ToF cameras, describing advantages, limitations and applications. It includes 68 references grouped in scene-related, object-related and human-related. Visual human-machine interaction is deeply studied as ToF cameras can be used for gesture recognition. Focusing on software, [113] reviews vision-based control algorithms for robot manipulators and [114] analyzes stereo matching algorithms used in robotics in terms of speed, accuracy, coverage, time consumption, and disparity range. Implementations of stereo matching algorithms in hardware for real-time applications are also discussed.

Table 4. Common applications in robotics.

	Scene-Related		**Object-Related**		
	People detection	Environment reconstruction / navigation	Object reconstruction / inspection	Bin picking / object manipulation	Robot pose / calibration
Stereo Vision and Photogrammetry			[74]		[18–20,67–73]
Projected Texture Stereo Vision				[76–78]	
Time of Flight	[95]	[79–87]	[88–91]	[92–94]	
Structured Light					[11,54,96]
Light Coding	[61,98–100]				
Laser Triangulation		[103–111]	[102]	[101]	[107]

Robots need flexibility and accuracy to carry out more complex and diverse tasks, such as collision avoidance with static and moving objects during navigation, collaborative work with humans, fine positioning, inspection, *etc.* In all the reviewed applications, each vision system has its single purpose. There has not been found one single vision system able to perform several tasks. Multiple vision systems are used (one for each individual purpose) instead of one single multi-tasking vision system. This is because requirements of each task are quite different and each technique has its scope and is more adequate than others.

5. CONCLUSIONS

In this survey, 3D vision solutions used in robotics have been organized, classified and reviewed. The purpose was to provide a compilation of the state of the art and the existing techniques so that future researchers and developers have a background information. Vision techniques for robot guidance have been analysed in terms of accuracy, range, weight, safety, processing time, and scanning environmental influences. Main advantages and main drawbacks have been also presented for each of them. Choosing which type of 3D vision system to use is highly dependent on the parts that need to be located or measured. While laser range finders using time of flight methods can be used to locate distant objects, stereo imaging systems may be better suited to imaging high-contrast objects. Where such objects are highly specular or textureless, it may be more useful to employ projected texture techniques. In addition to this, robot and industrial environments conditions also need to be considered. Each application and each type of robot need a specific vision solution. There is no universal vision technique to perform several tasks. Future woks may focus on multi-tasking or multi-purpose vision systems and their integration with other sensor types and systems.

Robots have become a core element of Industry 4.0 and flexibility can be incorporated to them by vision systems and other sensor technologies in order to achieve the requirements and functionalities of the new applications. New tasks are becoming more or more complex and it is necessary to improve the accuracy and to work collaborative with humans, which means making decisions in real-time and triggering actions. For these goals, visual feedback is the key issue, and this is in fact what vision systems provide to robots. Thus, 3D machine vision is the future for robotics. The idea of considering robot technology as an integral part of production is not new but nevertheless it is a challenge. Whether robots will be able to or should perform all the production steps in future is perhaps less a question of time than of money.

ACKNOWLEDGMENTS

The work was supported by European Union H2020 Programme under grant agreement No. 637107, SYMBIO-TIC.

AUTHOR CONTRIBUTIONS

All authors contributed extensively to the work presented in this paper and wrote the manuscript. Luis Pérez reviewed the state-of-the-art and wrote the initial version of the paper analyzing the vision techniques and their applications for robot guidance. Íñigo Rodríguez and Nuria Rodríguez participated with Luis in the analysis of the vision techniques. Rubén Usamentiaga and Daniel F. García provided suggestions and corrections during the preparation of the submitted paper and the review process.

CONFLICTS OF INTEREST

The authors declare no conflict of interest.

REFERENCES

1. Deane, P.M. *The First Industrial Revolution*; Cambridge University Press: Cambridge, UK, 1979.

2. Kanji, G.K. Total quality management: the second industrial revolution. *Total Qual. Manag. Bus. Excell.* **1990**, *1*, 3–12.

3. Rifkin, J. The third industrial revolution. *Eng. Technol.* **2008**, *3*, 26–27.

4. Kagermann, H.; Wahlster, W.; Helbig, J. *Recommendations for Implementing the Strategic Initiative Industrie 4.0: Final Report of the Industrie 4.0 Working Group*; Forschungsunion: Berlin, Germany, 2013.

5. Koeppe, R. New industrial robotics: human and robot collaboration for the factory. In Proceedings of the 2014 European Conference on Leading Enabling Technologies for Societal Challenges (LET'S 2014), Bologna, Italy, 29 September–1 October 2014.

6. European Commission. Factories of the Future in H2020. Available online: http://ec.europa.eu/research/industrial_technologies/ factories-of-the-future_en.html (accessed on 5 October 2015).

7. European Factories of the Future Research Association. *Factories of the Future: Multi-Annual Roadmap for the Contractual PPP under Horizon 2020*; Publications office of the European Union: Brussels, Belgium, 2013. [Google Scholar]

8. European Commission and Robotics. Available online: http://ec.europa.eu/programmes/horizon2020/en/h2020-section/robotics (accessed on 11 October 2015).

9. TAPAS Project. Available online: http://www.tapas-project.eu (accessed on 11 October 2015).

10. Bogh, S.; Schou, C.; Rühr, T.; Kogan, Y.; Dömel, A.; Brucke, M.; Eberst, C.; Tornese, R.; Sprunk, C.; Tipaldi, G.D.; Hennessy, T. Integration and assessment of multiple mobile manipulators in a real-world industrial production facility. In Proceedings of the 45th International Symposium on Robotics (ISR 2014), Munich, Germany, 2–4 June 2014; pp. 1–8.

11. VALERI Project. Available online: http://www.valeri-project.eu (accessed on 11 October 2015).

12. Zhou, K.; Ebenhofer, G.; Eitzinger, C.; Zimmermann, U.; Walter, C.; Saenz, J.; Pérez, L.; Fernández, M.A.; Navarro, J. Mobile manipulator is coming to aerospace manufacturing industry. In Proceedings of the 2014 IEEE International Symposium on Robotic and Sensors Environments (ROSE 2014), Timisoara, Romania, 16–18 October 2014; pp. 94–99.

13. SYMBIO-TIC Project. Available online: http://www.symbio-tic.eu (accessed on 11 October 2015).

14. Labudzki, R.; Legutko, S. Applications of Machine Vision. *Manuf. Ind. Eng.* **2011**, *2*, 27–29.

15. Wöhler, C. *3D Computer Vision: Efficient Methods and Applications*; Springer: Dortmund, Germany, 2009.

16. Alenyà, G.; Foix, S.; Torras, C. ToF cameras for active vision in robotics. *Sens. Actuators A Phys.* **2014**, *218*, 10–22.

17. Zhang, J.Y.; Zhao, C.; Zhang, D.W. Pose accuracy analysis of robot manipulators based on kinematics. *Adv. Mater. Res.***2011**, *201*, 1867–1872.

18. Hefele, J.; Brenner, C. Robot pose correction using photogrammetric tracking. In Proceedings of Intelligent Systems and Smart Manufacturing; International Society for Optics and Photonics, Boston, MA, USA, 12 February 2001; pp. 170–178.

19. Maas, H.G. Dynamic photogrammetric calibration of industrial robots. In Proceedings of Camera and System Calibration, San Diego, CA, USA, 27 July 1997.

20. Clarke, T.; Wang, X. The control of a robot end-effector using photogrammetry. *Int. Arch. Photogramm. Remote Sens.***2000**, *33*, 137–142.

21. Salvi, J.; Armangué, X.; Batlle, J. A comparative review of camera calibrating methods with accuracy evaluation.*Pattern Recognit.* **2002**, *35*, 1617–1635.

22. Faugeras, O. *Three-dimensional Computer Vision: A Geometric Viewpoint*; MIT Press: Massachusetts, MA, USA, 1993.

23. Usamentiaga, R.; Molleda, J.; García, D.F. Structured-light sensor using two laser stripes for 3D reconstruction without vibrations. *Sensors* **2014**, *14*, 20041–20063.

24. Sturm, P.; Ramalingam, S.; Tardif, J.P.; Gasparini, S.; Barreto, J. Camera models and fundamental concepts used in geometric computer vision. *Found. Trends Comput. Gr. Vis.* **2011**, *6*, 1–183.

25. Hanning, T. *High Precision Camera Calibration*; Springer: Wiesbaden, Germany, 2011.

26. Usamentiaga, R.; Molleda, J.; García, D.F.; Pérez, L.; Vecino, G. Real-time line scan extraction from infrared images using the wedge method in industrial environments. *J. Electron. Imaging* **2010**, *19*, 043017.

27. Wilson, A. Choosing a 3D Vision System for Automated Robotics Applications. *Vis. Syst. Des.* **2014**, *19*. Available online: http://www.vision-systems.com/articles/print/volume-19/issue-11/ features/choosing-a-3d-vision-system-for-automated-robotics-applications.html (accessed on 11 December 2014).

28. Ramos, B.; Santos, E. Comparative study of different digitization techniques and their accuracy. *Comput.-Aided Des.***2011**, *43*, 188–206. [

29. Yilmaz, H.M.; Yakar, M.; Gilec, S.A.; Dulgerler, O.N. Importance of digital close-range photogrammetry in documentation of cultural heritage. *J. Cult. Herit.* **2007**, *8*, 428–433.

30. Werner, T.; Zisserman, A. New techniques for automated architecture reconstruction from photographs. In Proceedings of the Seventh European Conference on Computer Vision (ECCV 2002), Copenhagen, Denmark, 28–31 May 2002; pp. 541–555.

31. Werner, T.; Zisserman, A. Model selection for automated architectural reconstruction from multiple views. In Proceedings of the Thirteenth British Machine Vision Conference (BMVC 2002), Cardiff, UK, 2–5 September 2002; pp. 53–62.

32. Werner, T.; Schaffalitzky, F.; Zisserman, A. Automated architecture reconstruction from close-range photogrammetry.*Int. Arch. Photogramm. Remote Sens. Spat. Inf. Sci.* **2002**, *34*, 352–359.

33. Lane, S.N.; James, T.D.; Crowell, M.D. Application of digital photogrammetry to complex topography for geomorphological research. *Photogramm. Rec.* **2000**, *16*, 793–821.

34. Fonstad, M.A.; Dietrich, J.T.; Corville, B.C.; Jensen, J.L.; Carbonneau, P.E. Topographic structure from motion: A new development in photogrammetric measurement. *Earth Surf. Process. Landf.* **2013**, *38*, 421–430.

35. Dueholm, K.S.; Garde, A.A.; Pedersen, A.K. Preparation of accurate geological and structural maps, cross-sections or block diagrams from

colour slides, using multi-model photogrammetry. *J. Struct. Geol.* **1993**, *15*, 933–937.

36. Eisenbeiss, H.; Lambers, K.; Sauerbier, M.; Li, Z. Photogrammetric documentation of an archaeological site using and autonomous model helicopter. *Int. Arch. Photogramm. Remote Sens. Spat. Inf. Sci.* **2005**, *36*, 238–243.

37. Granshaw, S.I. Bundle adjustment methods in engineering photogrammetry. *Photogramm. Rec.* **1980**, *10*, 181–207.

38. Fraser, C.S.; Brown, D.C. Industrial photogrammetry: New developments and recent applications. *Photogramm. Rec.***1986**, *12*, 197–217.

39. Goldan, M.; Kroon, R. As-built products modelling and reverse engineering in shipbuilding through combined digital photogrammetry and CAD/CAM technology. *J. Sh. Prod.* **2003**, *19*, 98–104.

40. Luhmann, T.; Robson, S.; Kyle, S.; Harley, I. *Close Range Photogrammetry. Principles, techniques and applications*; Whittles Publishing: Caithness, UK, 2006.

41. Canny, J. A computational approach to edge detection. *IEEE Trans. Pattern Anal. Mach. Intell.* **1986**, *8*, 679–698.

42. Harris, C.; Stephens, M. A combined corner and edge detector. In Proceedings of the Alvey vision Conference, Manchester, UK, 2 September 1998.

43. Lucas, B.D.; Kanade, T. An iterative image registration technique with an application to stereo vision. In Proceedings of the Seventh International Joint Conference on Artificial Intelligence (IJCAI 1981), Vancouver, BC, Canada, 24–28 August 1981; pp. 674–679.

44. Lowe, D.G. Distinctive image features from scale-invariant keypoints. *Int. J. Comput. Vis.* **2004**, *60*, 91–110.

45. Bay, H.; Ess, A.; Tuytelaars, T.; Van Gool, L. Speeded-up robust features (SURF). *Comput. Vis. Image Underst.* **2008**, *110*, 346–359.

46. Kimmel, R.; Zhang, C.; Bronstein, A.; Bronstein, M. Are MSER features really interesting? *IEEE Trans. Pattern Anal. Mach. Intell.* **2011**, *33*, 2316–2320.

47. IDS Whitepapers. Available online: https://en.ids-imaging. com/whitepaper.html (accessed on 30 July 2015).

48. Ensenso—Stereo 3D Cameras. Available online: http://www. ensenso.com (accessed on 13 December 2015).

49. Lim, J. Optimized projection pattern supplementing stereo systems. In Proceedings of the 2009 IEEE International Conference on Robotics and Automation (ICRA 2009), Kobe, Japan, 12–17 May 2009; pp. 2823–2829.

50. Konolige, K. Projected texture stereo. In Proceedings of the 2010 IEEE International Conference on Robotics and Automation (ICRA 2010), Anchorage, KY, USA, 3–7 May 2010; pp. 148–155.

51. Time of Flight-Heptagon. Available online: http://hptg.com/technology/time-of-flight/ (accessed on 13 December 2015).

52. Kolb, A.; Barth, E.; Koch, R.; Larsen, R. Time-of-flight sensors in computer graphics (state-of-the-art report). In Proceedings of Eurographics 2009-State of the Art Reports, Munich, Germany, 30 March–2 April 2009.

53. Chen, F.; Brown, G.; Song, M. Overview of three dimensional shape measurement using optical methods. *Opt. Eng.***2000**, *39*, 10–22.

54. Pagès, J.; Collewet, C.; Chaumette, F.; Salvi, J. A camera-projector system for robot positioning by visual servoing. In Proceedings of the 2006 Conference on Computer Vision and Pattern Recognition Workshop (CVPRW 2006), New York, NY, USA, 17–22 June 2006.

55. Salvi, J. An Approach to Coded Structured Light to Obtain Three Dimensional Information. Ph.D. Thesis, University of Gerona, Gerona, Spain, 1997.

56. Bernal, C.; Agustina, B.; Marín, M.M.; Camacho, A.M. Performance evaluation of optical scanner based on blue led structured light. *Proc. Eng.* **2013**, *63*, 591–598.

57. LMI Technologies. Available online: http://lmi3d.com/products (accessed on 30 October 2015).

58. Viager, M. *Analysis of Kinect for Mobile Robots*; Individual course report; Technical University of Denmark: Kongens Lyngby, Denmark, 2011.

59. Khoshelham, K.; Elberink, S.O. Accuracy and resolution of kinect depth data for indoor mapping applications. *Sensors***2012**, *12*, 1437–1454.

60. Mahmud, M.; Joannic, D.; Roy, M.; Isheil, A.; Fontaine, J.F. 3D part inspection path planning of a laser scanner with control on the uncertainty. *Comput.-Aided Des.* **2011**, *43*, 345–355.

61. Salmi, T.; Väätäinen, O.; Malm, T.; Montonen, J.; Marstio, I. Meeting new challenges and possibilities with modern robot safety technologies. In *Enabling Manufacturing Competitiveness and Economic Sustainability*; Springer International Publishing: Montreal, QC, Canada, 2014; pp. 183–188. [Google Scholar]

62. Point Cloud Library (PCL). Available online: http://pointclouds. org/ documentation/tutorials/ (accessed on 29 July 2015).

63. Luhmann, T. Close range photogrammetry for industrial applications. *ISPRS J. Photogramm. Remote Sens.* **2010**, *65*, 558–569.

64. Rieke-Zapp, D.; Tecklenburg, W.; Peipe, J.; Hastedt, H.; Haig, C. Evaluation of the geometric stability and the accuracy potential of digital cameras-Comparing mechanical stabilisation versus parametrisation. *ISPRS J. Photogramm. Remote Sens.* **2009**, *64*, 248–258.

65. Broers, H.; Jansing, N. How precise is navigation for minimally invasive surgery? *Int. Orthop.* **2007**, *31*, 39–42.

66. Laser Tracker Systems—Leica Geosystems. Available online: http:// www.leica-geosystems.com/en/Laser-Tracker-Systems_69045.htm (accessed on 2 November 2015).

67. Shirinzadeh, B.; Teoh, P.L.; Tian, Y.; Dalvand, M.M.; Zhong, Y.; Liaw, H.C. Laser interferometry-based guidance methodology for high precision positioning of mechanisms and robots. *Robotics Computer-Integrated Manuf.* **2010**, *26*, 74–82.

68. Vincze, M.; Prenninger, J.P.; Gander, H. A laser tracking system to measure position and orientation of robot end effectors under motion. *Int. J. Robotics Res.* **1994**, *13*, 305–314.

69. Dumas, C.; Caro, S.; Mehdi, C.; Garnier, S.; Furet, B. Joint stiffness identification of industrial serial robots. *Robotica***2012**, *30*, 649–659.

70. Qu, W.W.; Dong, H.Y.; Ke, Y.L. Pose accuracy compensation technology in robot-aided aircraft assembly drilling process. *Acta Aeronaut. Astronaut. Sinica* **2011**, *32*, 1951–1960.

71. Leikas, E. Robot guidance with a photogrammetric 3-D measuring system. *Ind. Robot* **1999**, *26*, 105–108.

72. Hefele, J.; Brenner, C. Real-time photogrammetric algorithms for robot calibration. *Int. Archives Photogramm. Remote Sens. Spat. Inf. Sci.* **2002**, *34*, 33–38.

73. Liu, B.; Zhang, F.; Qu, X. A method for improving the pose accuracy of a robot manipulator based on multi-sensor combined measurement and data fusion. *Sensors.* **2015**, *15*, 7933–7952.

74. Amdal, K. Single camera system for close range industrial photogrammetry. *Int. Archives Photogramm. Remote Sens.***1992**, *29*, 6–10.

75. Luhmann, T. Precision potential of photogrammetric 6DOF pose estimation with a single camera. *ISPRS J. Photogramm. Remote Sens.* **2009**, *64*, 275–284.

76. IDS Case Studies. Available online: https://en.ids-imaging.com/case-studies.html (accessed on 30 July 2015).

77. Carroll, J. 3D vision system assists in robotic bin picking. *Vis. Syst. Des.* **2014**, *19*. Available online: http://www.vision-systems.com/ articles/ 2014/08/3d-vision-system-assists-in-robotic-bin-picking.html (accessed on 29 August 2014).

78. Sturm, J.; Konolige, K.; Stachniss, C.; Burgard, W. 3D pose estimation, tracking and model learning of articulated objects from dense depth video using projected texture stereo. In Proceedings of the 2010 RGB-D: Advanced Reasoning with Depth Cameras Workshop, Zaragoza, Spain, 27 June 2010.

79. Weingarten, J.W.; Gruener, G.; Siegwart, R. A state-of-the-art 3D sensor for robot navigation. In Proceedings of the 2004 IEEE/RSJ International Conference on Intelligent Robots and Systems (IROS 2004), Sendai, Japan, 28 September–2 October 2004; pp. 2155–2160.

80. May, S.; Werner, B.; Surmann, H.; Pervolz, K. 3D time-of-flight cameras for mobile robotics. In Proceedings of the 2006 IEEE/RSJ International Conference on Intelligent Robots and Systems (IROS 2006), Beijing, China, 9–15 October 2006; pp. 790–795.

81. May, S.; Droeschel, D.; Holz, D.; Wiesen, C.; Fuchs, S. 3D pose estimation and mapping with time-of-flight cameras. In Proceedings of the 2008 IEEE/RSJ International Conference on Intelligent Robots and Systems (IROS 2008), 3D Mapping Workshop, Nice, France, 22–26 September 2008.

82. Hedge, G.; Ye, C. Extraction of planar features from Swissranger SR-3000 range images by a clustering method using normalized cuts. In Proceedings of the 2009 IEEE/RSJ International Conference on Intelligent Robots and Systems (IROS 2009), St. Louis, MO, USA, 10–15 October 2009; pp. 4034–4039.

83. Ohno, K.; Nomura, T.; Tadokoro, S. Real-time robot trajectory estimation and 3D map construction using 3D camera. In Proceedings of the 2006 IEEE/RSJ International Conference on Intelligent Robots and Systems (IROS 2006), Beijing, China, 9–15 October 2006; pp. 5279–5285.

84. Stipes, J.A.; Cole, J.G.P.; Humphreys, J. 4D scan registration with the SR-3000 LIDAR. In Proceedings of the 2008 International Conference on Robotics and Automation (ICRA 2008), Pasadena, CA, USA, 19–23 May 2008; pp. 2988–2993.

85. Arbeiter, G.; Fischer, J.; Verl, A. 3-D-Environment reconstruction for mobile robots using fast-SLAM and feature extraction. In Proceedings of the Forty-first International Symposium on Robotics (ISR 2010), Munich, Germany, 7–9 June 2010.

86. Kuhnert, K.D.; Stommel, M. Fusion of stereo-camera and PMD-camera data for realtime suited precise 3D environment reconstruction. In Proceedings of the 2006 IEEE/RSJ International Conference on Intelligent Robots and Systems (IROS 2006), Beijing, China, 9–15 October 2006; pp. 4780–4785.

87. Netramai, C.; Oleksandr, M.; Joochim, C.; Roth, H. Motion estimation of a mobile robot using different types of 3D sensors. In Proceedings of the Fourth International Conference on Autonomic and Autonomous Systems (ICAS 2008), Gosier, France, 16–21 March 2008; pp. 148–153.

88. Dellen, B.; Alenyà, G.; Foix, S.; Torras, C. 3D object reconstruction from swissranger sensors data using a spring-mass model. In Proceedings of the Fourth International Conference on Computer Vision Theory Applications, Lisbon, Portugal, 5–8 February 2009; pp. 368–372.

89. Foix, S.; Alenyà, G.; Andrade-Cetto, J.; Torras, C. Object modelling using a ToF camera under an uncertainty reduction approach. In Proceedings of the 2010 IEEE International Conference on Robotics Automation (ICRA 2010), Anchorage, KY, USA, 3–7 May 2010; pp. 1306–1312.

90. Haddadin, S.; Suppa, M.; Fuchs, S.; Bodenmüller, T.; Albu-Schäffer, A.; Hirzinger, G. Towards the robotic co-worker.*Robotics Res.* **2011**, *70*, 261–282.

91. Fuchs, S.; May, S. Calibration and registration for precise surface reconstruction with time of flight cameras. *Int. J. Intell. Syst. Technol. Appl.* **2008**, *5*, 274–284. [Google Scholar] [CrossRef]

92. Kuehnle, J.U.; Xue, Z.; Stotz, M.; Zoellner, J.M.; Verl, A.; Dillmann, R. Grasping in depth maps of time-of-flight cameras. In Proceedings of the 2008 IEEE International Workshop on Robotic and Sensors Fnvironments (ROSE 2008), Ottawa, ON, Canada, 17–18 October 2008; pp. 132–137.

93. Saxena, A.; Wong, L.; Ng, A.Y. Learning grasp strategies with partial shape information. In Proceedings of the Twenty-third AAAI Conference on Artificial Intelligence, Chicago, IL, USA, 13–17 July 2008; pp. 1491–1494.

94. Maldonado, A.; Klank, U.; Beetz, M. Robotic grasping of unmodeled objects using time-of-flight range data and finger torque information. In Proceedings of the 2010 IEEE/RSJ International Conference on Intelligent Robots and Systems (IROS 2010), Taipei, Taiwan, 18–22 October 2010; pp. 2586–2591.

95. Foix, S.; Alenyà, G.; Torras, C. *Exploitation of Time-of-Flight (ToF) Cameras*; Technical Report; CSIC-UPC: Barcelona, Spain; December; 2010.

96. Claes, K.; Bruyninckx, H. Robot positioning using structured light patterns suitable for self calibration and 3D tracking. In Proceedings of the Thirteenth International Conference on Advanced Robotics, Daegu, Korea (South), 22–25 August 2007.

97. Le Moigne, J.J.; Waxman, A.M. Structured light patterns for robot mobility. *J. Robotics Autom.* **1988**, *4*, 541–548.

98. Patra, S.; Bhowmick, B.; Banerjee, S.; Kalra, P. High resolution point cloud generation from kinect and HD cameras using graph cut. In Proceedings of the 2012 International Joint Conference on Computer Vision, Imaging and Computer Graphics Theory and Applications (VISAPP 2012), Rome, Italy, 24–26 February 2012.

99. Susperregi, L.; Sierra, B.; Castrillón, M.; Lorenzo, J.; Martínez-Otzeta, J.M.; Lazkano, E. On the use of a low-cost thermal sensor to improve kinect people detection in a mobile robot. *Sensors.* **2013**, *13*, 14687–14713.

100. Wang, L.; Schmidt, B.; Nee, A.Y.C. Vision-guided active collision avoidance for human-robot collaborations. *Manuf. Lett.* **2013**, *1*, 5–8.

101. FlexSort Project. Available online: http://www.prodintec. es/ prodintec/g_noticias?accion=detalleNoticia&id=159(accessed on 15 November 2015).

102. Brosed, F.J.; Santolaria, J.; Aguilar, J.J.; Guillomía, D. Laser triangulation sensor and six axes anthropomorphic robot manipulator modelling for the measurement of complex geometry products. *Robot. Comput.-Integr. Manuf.* **2012**, *2*, 660–671.

103. Nguyen, V.; Martinelli, A.; Tomatis, N.; Siegwart, R. A comparison of line extraction algorithms using 2D laser rangefinder for indoor mobile robotics. In Proceedings of the 2005 IEEE/RSJ International Conference on Intelligent Robots and Systems (IROS 2005), Edmonton, AB, Canada, 2–6 August 2005; pp. 1929–1934.

104. Borges, G.A.; Aldon, M.J. line extraction in 2D range images for mobile robotics. *J. Intell. Robotic Syst.* **2004**, *40*, 267–297.

105. Agapakis, J.E.; Katz, J.M.; Friedman, J.M.; Epstein, G.N. Vision-aided robotic welding: An approach and a flexible implementation. *Int. J. Robotics Res.* **1990**, *9*, 17–34.

106. Fernández, A.; Acevedo, R.G.; Alvarez, E.A.; López, A.C.; García, D.F.; Usamentiaga, R.; Sánchez, J. Low-cost system for weld tracking based on artificial vision. *IEEE Trans. Ind. Appl.* **2011**, *47*, 1159–1167.

107. Khadraoui, D.; Motyl, G.; Martinet, P.; Gallice, J.; Chaumette, F. Visual servoing in robotics scheme using a camera/laser-stripe sensor. In Proceedings of the 1996 IEEE International Conference on Robotics and Automation (ICRA 1996), Minneapolis, MN, USA, 22–28 April 1996; pp. 743–750.

108. Pears, N.; Probert, P. An optical range sensor for mobile robot guidance. In Proceedings of the 1993 IEEE International Conference on Robotics and Automation (ICRA 1993), Atlanta, GA, USA, 2–6 May 1993; pp. 659–664.

109. Einsele, T. Real-time self-localization in unknown indoor environments using a panorama laser range finder. In Proceedings of the 1997 IEEE/RSJ International Conference on Intelligent Robots and Systems (IROS 1997), Grenoble, France, 11 September 1997; pp. 697–702.

110. Jensfelt, P.; Christensen, H. Laser based position acquisition and tracking in an indoor environment. In Proceedings of the 1998 International Symposium on Robotics and Automation (ISRA 1998), Leuven, Belgium, 20 May 1998.

111. Zhang, L.; Ghosh, B.K. Line segment based map building and localization using 2D laser rangefinder. In Proceedings of the 2000 IEEE International Conference on Robotics and Automation (ICRA 2000), San Francisco, CA, USA, 24–28 April 2000; pp. 2538–2543.

112. Gühring, J. Dense 3D surface acquisition by structured light using off-the-shelf components. In Proceedings of Videometrics and Optical Methods for 3D Shape Measurement, San Jose, CA, USA, 20 January 2001. SPIE: 2001.

113. Hashimoto, K. A Review on vision-based control of robot manipulators. *Adv. Robotics* **2003**, *17*, 969–991.

114. Lazaros, N.; Sirakoulis, G.C.; Gasteratos, A. review of stereo vision algorithms: from software to hardware. *Int. J. Optomechatronics* **2008**, *2*, 435–462.]

Chapter 7

An Imaging Sensor-Aided Vision Navigation Approach that Uses a Geo-Referenced Image Database

Yan Li [1], **Qingwu Hu** [2,*,†], **Meng Wu** [3,†] **and Yang Gao** [3,†]

[1] School of Information Engineering, Chang'an University, Xi'an 710064, China

[2] School of Remote Sensing and Information Engineering, Wuhan University, Wuhan 430079, China

[3] Xi'an Research Institute of Surveying and Mapping, Xi'an 710054, China

ABSTRACT

In determining position and attitude, vision navigation via real-time image processing of data collected from imaging sensors is advanced without a high-performance global positioning system (GPS) and an inertial measurement unit (IMU). Vision navigation is widely used in indoor navigation, far space navigation, and multiple sensor-integrated mobile mapping. This paper proposes a novel vision navigation approach aided by imaging sensors and that uses a high-accuracy geo-referenced image database (GRID) for high-precision navigation of multiple sensor platforms in environments with poor GPS. First, the framework of GRID-aided vision navigation is developed with sequence images from land-based mobile mapping systems that integrate multiple sensors. Second, a highly efficient

GRID storage management model is established based on the linear index of a road segment for fast image searches and retrieval. Third, a robust image matching algorithm is presented to search and match a real-time image with the GRID. Subsequently, the image matched with the real-time scene is considered to calculate the 3D navigation parameter of multiple sensor platforms. Experimental results show that the proposed approach retrieves images efficiently and has navigation accuracies of 1.2 m in a plane and 1.8 m in height under GPS loss in 5 min and within 1500 m.

KEYWORDS

vision navigation; imaging sensor; geo-referenced; image database; multiple sensor-integrated mobile mapping; image retrieval; image matching

1. INTRODUCTION

The objective of a moving platform is to provide accurate information for moving platforms, e.g., a land vehicle, an airplane, or aircraft in space, particularly when a global positioning system (GPS) or an inertial measurement unit (IMU) is not working effectively under poor environmental conditions [1,2]. Imaging sensors have been widely studied and applied to determine position and orientation via a technology called vision navigation. This approach may be effective for such applications because it can be utilized in a GPS/IMU-deprived environment (such as indoors, in urban canyons, and in far space) [3,4,5]. Accurate positioning and orientation in a downtown building forest district or a long-distance tunnel with GPS outage and longtime IMU drift is a significant challenge in 3D navigation of a multiple sensor-integrated platform even with high-performance GPS/IMU-integrated platforms, such as the land-based mobile mapping system (L-MMS) for 3D surveying and modeling. Imaging sensor-based vision navigation can avoid certain harsh environmental restrictions because this approach does not depend on any signal or radiant sources [1,6]. Nonetheless, this environmental condition becomes problematic when real-time and high-precision performance is required.

Imaging sensor-based vision navigation estimates position and orientation information according to the geometrical relation from overlapping sequence images, which can improve the reliability of a navigation platform. DeSouza and Kak [7] investigated the developments in vision-related fields for mobile robot navigation over the past 20 years. Vision navigation is classified into three different categories: map-based, map building-based, and mapless navigation [7,8,9]. Mapless vision navigation utilizes imaging sensors and does not consider any prior description of the environment. Sequence images from the imaging sensors are used for motion analysis to determine the relative position and orientation information of a moving platform. The recently developed simultaneous localization and mapping (SLAM) technique integrates imaging sensors to extract the 3D navigation information with which 3D environment data are used for a map building-based vision navigation application. SLAM is employed in self-driving cars, unmanned aerial vehicles, autonomous underwater vehicles, planetary rovers, new domestic robots, and even within the human body [10,11,12]. At present, imaging sensors are combined with other laser scanning sensors to obtain robust positions and orientations for SLAM-based applications. Mapless vision navigation is a relative navigation technology that supports moving platforms to understand surroundings and explore a local environment, such as in indoor navigation and self-driving cars. This type of navigation stores information from the current environment, as well as its own relative position in the environment.

Nonetheless, cases of absolute navigation with global 3D coordinates and orientation have been observed, such as outdoor navigation, the L-MMS for 3D surveying platforms, and space navigation [2,13,14]. These cases require global geo-referencing with highly accurate absolute 3D coordinates and may include GPS/IMU; however, harsh and poor environmental conditions limit the performance of GPS/IMU navigation. Thus, map-based vision navigation is implemented by providing a moving platform with a geo-referenced 3D model of the environment, such as a 3D map, landmark, and geo-referenced image database (GRID) [3,15,16]. Map-based vision navigation considers the map as a sensor to match the real-time imaging sensor and facilitate highly accurate absolute navigation in advanced driver assistance systems (ADAS). A collection of geo-referenced images serves as a map for real-time localization; this collection is normally treated as a 3D map because it contains not only a set of landmarks but also the corresponding 3D location information [17,18]. Thus, imaging sensor-based vision navigation has two application scenarios as a map-based navigation approach: one is to improve the reliability of navigation under harsh environmental conditions in which GPS and IMU are ineffective, such as in ADAS. The other is to improve navigation accuracy where the vision navigation result is considered to be new input for Kalman filtering with GPS/IMU, such as in L-MMS.

The key issues in image database-aided vision navigation are image searching and the matching methods with real-time images (RTIs) [19,20]. Results of recent research show that the process of matching two images of the same scene under different scales, illumination, and view angles has been developed successfully for years; nonetheless, several theoretical and technical problems must be addressed [21,22]. For example, two independent images share many similar features or pixels under certain circumstances, which may cause matching ambiguity given many candidate images. This condition is a serious problem for image database-based vision navigation. The speed of matching RTIs with the image database not only depends on feature computation but also on image search and retrieval. The organization model of the image database is combined with robust feature matching to realize matching with fast searching; these methods are the key techniques in GRID-based vision navigation.

Our paper presents a novel imaging sensor-aided vision navigation approach that uses the highly accurate GRID. First, the framework of GRID-aided vision navigation is established with sequence images derived from land-based, multiple sensor-integrated mobile mapping systems. Second, a highly efficient GRID storage management model is developed based on the linear index of a road segment for fast image search and retrieval. Third, a robust image matching algorithm is presented to search for and match RTIs with GRID; the matched image is then matched with the real-time scene to calculate the 3D navigation parameter of the multiple-sensor platform. Experimental results show that the proposed approach retrieves images highly efficiently and has navigation accuracies of 1.2 m in a plane and 1.8 m in height under GPS loss in 5 min and within 1500 m.

2. METHODOLOGY

2.1. Framework and Navigation Model of GRID-Aided Vision Navigation

The foundation of the proposed vision navigation approach is the comprehensive GRID with highly accurate position and orientation parameters that were derived from an L-MMS prior to vision navigation. All the images are captured and synchronized with the GPS/IMU according to image exposure time to calculate these parameters. The geo-referenced images are collected and are considered to be a highly accurate 3D survey of the environment. The imaging sensors of the moving platform capture the RTI of the current scene. Subsequently, this image is presented to match and search the GRID to obtain the exact geo-referenced image. Thus, the geo-referenced data are transferred to the RTI on the basis of the matched point with a strict photogrammetry geometry model to determine the 3D coordinates and the orientation parameter of the moving platform upon capturing the RTI. The framework of the proposed imaging sensor-aided vision navigation method with GRID is presented in Figure 1.

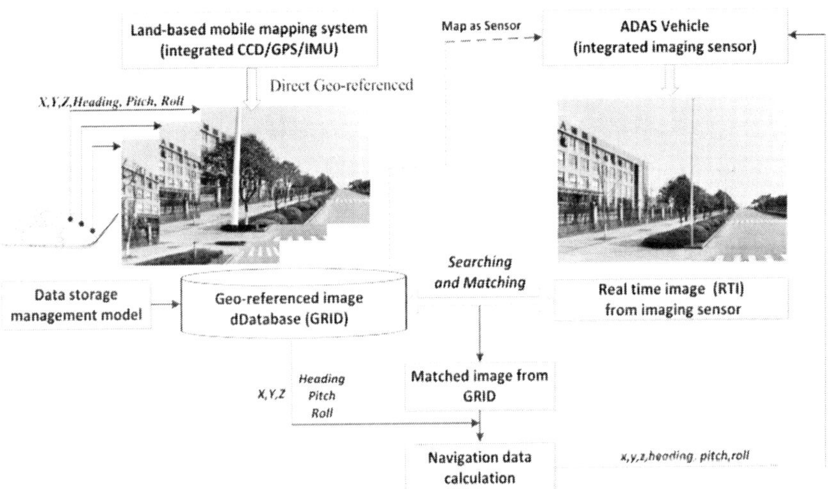

Figure 1. Framework of the proposed imaging sensor-aided vision navigation approach that uses a geo-referenced database.

Three key issues are observed in the proposed vision navigation approach in relation to feasibility and accuracy; one is the performance of the image database. The geo-referenced image should be indexed well to reduce the time consumed in querying a candidate image set from the entire image database. Thus, we propose a data storage management model for GRID, as introduced in Section 2.2. The other issue involves the fast image searching and matching of the candidate image set with the RTI. The two processes should be completed quickly with high reliability, as described in

Section 2.3. The third issue involves transferring the 3D navigation parameter from the geo-referenced image to the RTI precisely. We propose a vision navigation model from the image database to the moving platform, as explained in Section 2.4.

2.2. GRID Data Storage Management Model

2.2.1. L-MMS for GRID Data Collection

The GRID is compiled by multiple sensor-integrated, land-based mobile mapping systems, which were developed in the 1990s for rapid and efficient road surveying. Such systems include VISAT™, GPS Vision™, GI-EYE™, LD2000™, ON-SIGHT™, Topcon IP S2, and Hi-Target iScan [2,23,24,25]. L-MMS integrates geodetic quality GPS, an inertial navigation system (INS), an odometer, a synchronization time board, and digital cameras that are mounted on a land vehicle (as presented in Figure 2). As a result, highly accurate GPS data can be logged along with INS orientation data and sequence images with high synchronized time in the process of running along defined routes at a speed of no more than 80 km/h.

Four or more cameras are installed in L-MMS to capture sequence images with a user-defined interval distance from 1 to 10 m. All the images are tagged with a time label by the synchronization time board, which adopts the GPS time as a time clock with an accuracy of less than 0.01 ms. The GPS/INS provides the orientation parameters for each sequence image based on the unique and unified GPS time. L-MMS is the most rapid, convenient, accurate, and economical surveying system for collecting and updating geo-information together with the GRID.

One L-MMS can obtain geo-referenced images for more than 200 km. Each image with a resolution of more than 2 MB in RGB mode has a data volume of approximately 5 MB. If the interval distance in image capturing is 5 m, then the number of images at 200 km will be 160,000 and the data volume will be 800,000 MB [2,26]. Determining the image number and the data volume are difficulties encountered in searching and matching RTI and GRID. The strategy and method of GRID design affects the efficiency of image searching and matching. To ensure the efficiency of the proposed vision navigation, we propose the indexing of a large number of images according to dynamic road section and the packing of images based on a large file storage model.

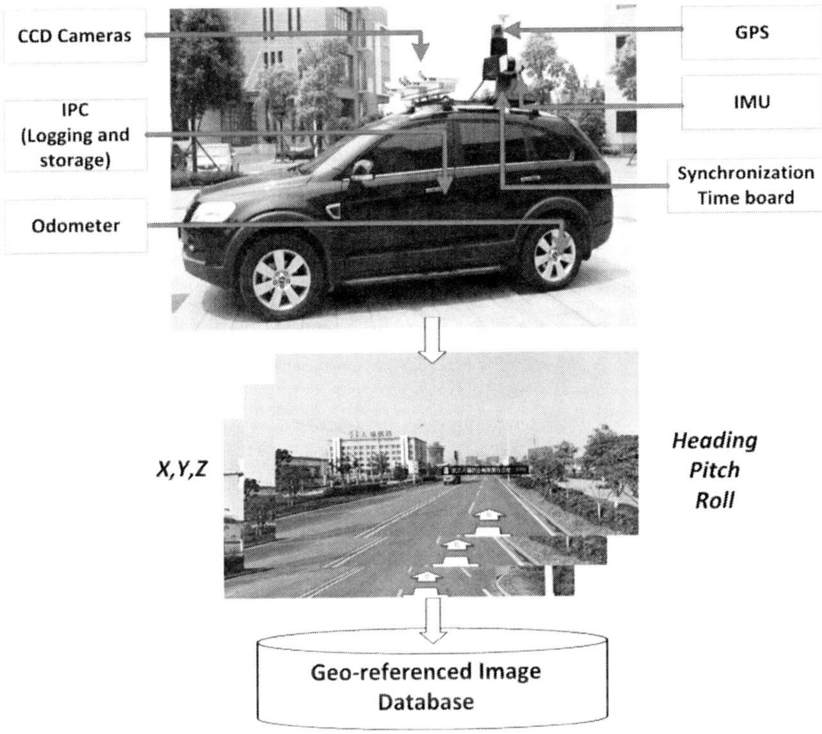

Figure 2. L-MMS for the data collection of geo-referenced database.

2.2.2. Dynamic Section Indexing of GRID Based on Road Networks

At present, the quad-tree index is commonly used as a data organization and indexing method (such as the KIWI data format developed in Japan) [27,28] to navigate digital maps. The basic principle of this method is based on a quad-tree structure, and the geospace is divided into small grids. The map elements in the bottom grid to which the quad-tree leaf nodes correspond are packaged and stored according to location and size; furthermore, the map data are spatially separated into small blocks. The main advantage of the quad-tree index is its capability to quickly locate the current target area for a specified grid (block). The relevant target feature can also be extracted from an entire block. Nonetheless, obvious drawbacks are detected in GRID data indexing. GRID data are usually sampled along roads; therefore, data distribution in a quad-tree grid is uneven. As a result, the index data are redundant. In addition, all the images within the grid must be traversed and compared one by one in single-image retrieval (depending on the shooting location of the image), thereby lowering retrieval efficiency. Therefore, the GRID data indexing model in particular must be examined and designed.

GRID data indexing must satisfy two conditions. First, the traveling vehicle should be capable of quickly retrieving a recent image in a trend direction before reaching the current positioning point. Second, this approach should be capable of checking the image sequence in which the current positioning point of the region may be distributed during image matching and location retrieval. GRID data are linearly distributed along a road, unlike 2D GIS data with the geometrical features of point, line, and surface. Given this feature, a road-based dynamic segmentation (DS) index structure is designed in the linear referencing system to determine the spatial query speed of GRID and to ensure quick image matching.

The linear referencing system (LRS) constitutes a method of recording linearly distributed target position information. This system is widely used in the transport domain [29,30,31,32]. LRS can be expressed as (*R, M*), which are also known as linear coordinates that indicate the position information of linear characteristics in an relative offset *M*, *i.e.*, the distance to the starting point of the linear feature. *R* denotes the linear characteristic. For example, *(320, 274)* suggests that the distance from the starting point is *274* at road *320*. Essentially, the proposed DS index method does not alter the position of image distribution (shooting position). Moreover, the image space coordinate in the 2D reference system is not associated with a road in the linear referencing system to establish the GRID data index depending on the road network. The specific approach adopts DS technology according to the image distribution position and location section, applies the offset from the starting point of each road section representing the image position, and converts the image position from 2D coordinates *(X, Y)* to linear coordinates (*M*). Meanwhile, road split (non-physical division) is realized, and the image is hooked with the corresponding road section. A DS index (DS) tree is also built into the linear benchmark for each linear feature (road).

The shape of the DS tree is similar to those of the *R* tree and quad-tree. However, the latter two indices are located in a 2D space, whereas the DS tree is indexed in a 1D linear space. In addition, the quad-tree and *R* tree represent the position of an object in an index space given the spatial coordinates of object (*X, Y*) or a minimum bounding rectangle, whereas the

DS tree indicates the position of an object through offset M of the object. This tree is composed of root, intermediate, and leaf nodes. The 1D data set is divided recursively according to the nodes in each layer. In Figure 3, the two-direction road *R* includes two road-width-flow centerlines *C* and *D* that correspond to two road section sequences in opposite directions: *C1, C2, C3*and *D1, D2, D3*, respectively. These sequences comprise image sequences *P1, P2, ..., P9* and *K1, K2, ..., K9*. Figure 3a presents a schematic of the division of road R via DS index intervals. Figure 3b illustrates the corresponding DS index tree structure, where R is the root node. The structure of R can be expressed as R(*RID, M1, M2*). RID is the logical road index number. *M1* and *M2* are the beginning and ending offsets of the entire road. The starting offset of the road (*M1*) is generally set to 0. Intermediate node *C* or *D* includes section sequences along two directions of the centerline of road way flow, and the related structure can be expressed as S(*SID, M1, M2*). *SID* is the section index number, and *M1* and *M2* are the beginning and ending offsets of the sections along road flow, respectively. The leaf node stores information on each image in the GRID image sequence, and its structure is *I (IID, M)*. *IID* is the image index number, and M is the offset of the sections in which the image position is detected.

The data structure of the index tree node of GRID data is presented as follows:

```
struct NODE
{
Linear_Interval linear;          //linear interval
NODE*Parent;                                //a pointer to the parent node
NODE**child;                   //a pointer to the child node
long*ObjID;                        //object ID within the linear interval
}
```

The quad-tree index in the 2D reference system must calculate the Euclidean distance between the location point and the image position individually for image retrieval. Meanwhile, the DS index in the linear referencing system only needs to compare the location point offset with the image position offset. The retrieval calculations are performed in a 1D space. Thus, the computation amount is reduced, and the retrieval efficiency can be improved.

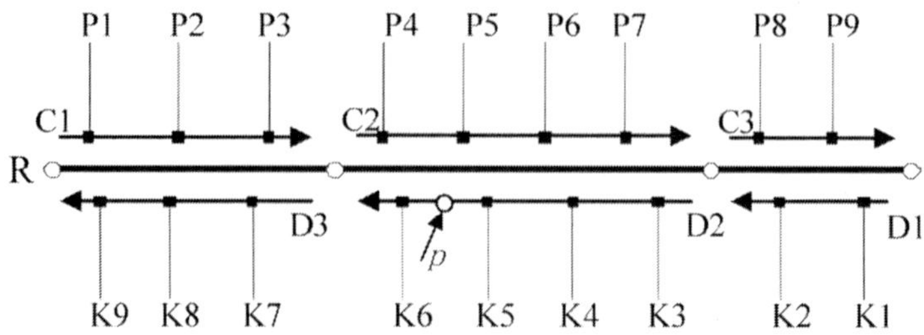

(a) DS Index Space Division

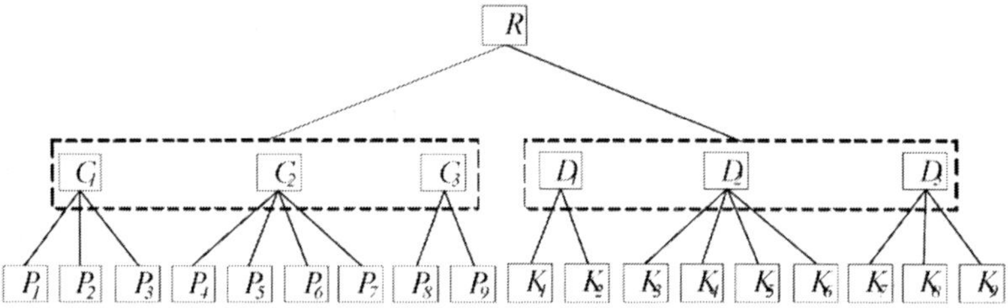

(b) DS Index Tree Structure

Figure 3. Dynamic segment indexing of GRID based on the road network LRS: (**a**) division of DS index space; (**b**) structure of the DS index tree.

2.2.3. GRID Data Storage Based on Large Files

Many GRID images are collected by L-MMS. If 1 site includes four images and the distance between the two sites is 5 m, then the GRID data comprise more than 800,000 images given a city with 1000-km roads. If each image corresponds to a data file, then an excessive amount of files is produced, thus directly affecting data extraction efficiency. In addition, many small files induce frequent disk reading and writing, thereby resulting in straightforward disk damage. Therefore, the design of the data storage model for GRID must be optimized.

To effectively control the number of the generated GRID data files, a data storage model based on large files is proposed in this paper. The basic idea involves considering one road section as a unit to centrally store the image sequence in the section. This concept not only reduces the number of data files but also quickly determines the image sequence through a road section through the DS index and improves the efficiency of image retrieval.

The organization of GRID data based on large files is line with the following rules:

Rule 1: Images in one road section should be stored in one large data file as much as possible;

Rule 2: Cross-section images should be stored separately;

Rule 3: If the images in a section are so numerous that they constitute an excessively large image file (exceeding the set threshold), then this image file should be split and stored as multiple files.

The proposed large file-based data storage model for GRID is established through compliance with the aforementioned rules, as shown in Figure 4.

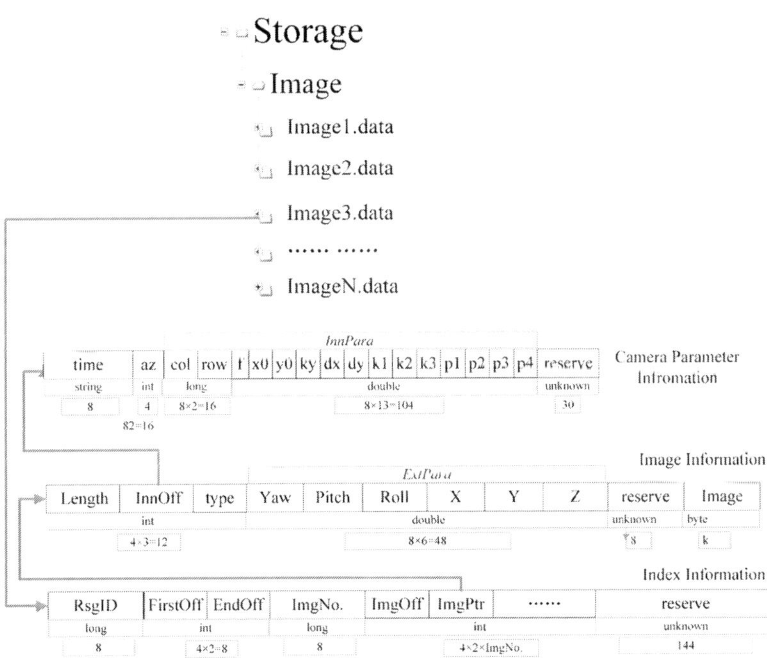

Figure 4. Large file-based data storage model for GRID.

A large file for GRID can be divided into three parts: an image index table, a file body, and an image parameter file. The index table records the current file section ID (long type, eight bytes), first section offset of the image position in the file (int type, four bytes), final section offset of the image position (int type, four bytes), number of images (long type, eight bytes), section offset of each next image in turn (int type, four bytes), and stored address of the image in the file (int type, four bytes). The file body accounts for each image information record (44 bytes plus the space occupied by the image sequence). The image parameter file stores the interior and exterior orientations of the GRID. The size of each large file for this database should be controlled to below 200 MB for one road section. When the file size exceeds this recommendation, a new large file should be generated automatically. One record should not be stored in two separate files. If the current storage space is not enough to store an entire piece of data, then this information should be transferred to a new large file for storage. All large GRID files are labeled "Image" + n + ". data".

The proposed GRID data storage based on large files can be organized in a database system or file directory, and this step is similar to image retrieval. In database storage, the generation of a large file can be regarded as the storage of sequence images related to a road section via a memory block. Meanwhile, the single file model can be treated as individual image storage.

2.3. Fast Image Searching and Matching

The RTI from a global GRID is impossible to search for and match efficiently or accurately. A navigation system typically has GPS and IMU sensors, which play the most important role in determining the position and orientation parameters of a moving platform. Imaging sensor-based vision navigation is adopted to enhance navigation reliability and accuracy for a moving platform in a harsh GPS or IMU-failed environment. Thus, the vision navigation module generates an initial navigation parameter from GPS/IMU. The initial position can restrict the range of GRID searching and matching, which can induce fast image searching and accurate matching. The proposed method to search and match the RTI from GRID includes three steps: searching for and locating large GRID files, extracting image features, and image matching.

2.3.1. Searching for and Locating Large GRID Files

To obtain a fast and reliable search result from a high-volume GRID, a spatial query is implemented before image retrieval based on image understanding. Although the worst position accuracy of GPS/IMU is 100–500 m, it can restrict image searching and matching in a road section or a street block. This restriction can significantly reduce the workload of blind searching and matching over a wide range. Thus, a fast search that conducts a spatial query of the initial position can be performed through the proposed dynamic section indexing of GRID based on a road network. The road section of the initial position can be determined to locate a certain large GRID file for accurate matching. Figure 5 illustrates the proposed concept of fast searching based on spatial query to locate a large GRID file.

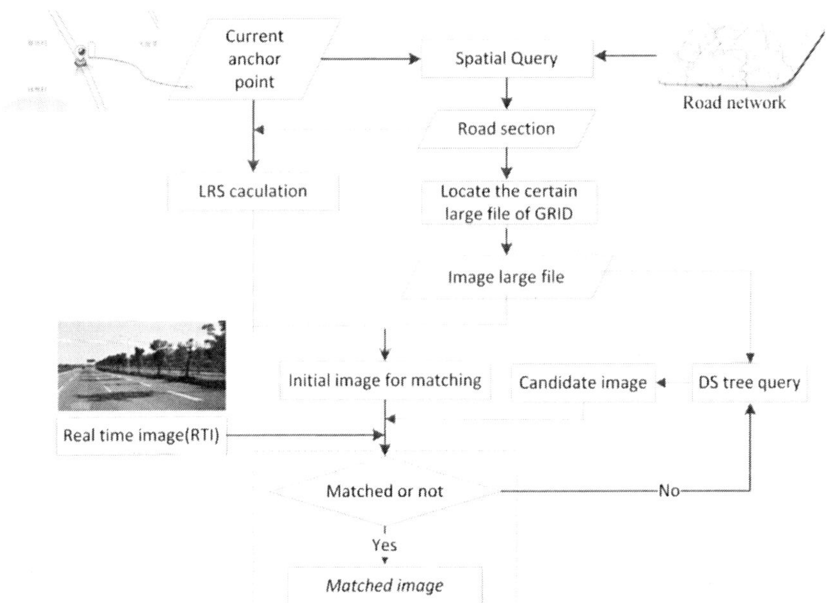

Figure 5. Fast search and large file location based on spatial query.

The current anchor point from GPS/IMU is considered as the spatial query input within the road network to determine the current road section. A large file can be located through the road section based on the proposed GRID data model. Meanwhile, the anchor point is converted into a 1D coordinate based on the LRS model to obtain the initial image from a large file as the candidate image for matching with the RTI. The candidate image is identified based on the DS tree; the process stops until the image is well matched with the RTI, as explained in the subsequent part.

2.3.2. Extracting Image Features

RTI and GRID are the images captured in different environments of the same scene. Feature selection and extraction are the key issues in achieving highly reliable image matching. Lowe [33,34] proposed an efficient scale invariant feature transform (SIFT) algorithm that is widely used in image matching. The local features of SIFT include invariance in rotation, scaling, affine transformation, and a view angle, which is suitable for matching images captured at different times, distances, and viewing angles [33,34,35,36]. In the proposed approach, we adopt a 128-dimensional SIFT feature vector to describe the key feature point. The geometric distortion influence of scale and rotation can be reduced by this vector, and its normalization can eliminate the effect of illumination change. Thus, the SIFT vector of an image local feature can result in robust matching between RTI and GRID.

The SIFT 128-dimensional feature vector in GRID is time consuming to extract, particularly in real-time processing applications. To decrease the duration of matching between RTI and GRID, the SIFT feature vectors of each image in GRID are extracted in advance and linked to GRID.

2.3.3. Image Matching

Image matching between RTI and GRID is implemented based on the SIFT feature. As this process runs during the image search, SIFT matching should be as fast as possible. The SIFT feature vectors of RTI can be extracted before matching because we have identified these vectors for each image in the GRID in advance. Thus, image matching is conducted only to calculate the similarity among the SIFT feature vectors of RTI and GRID. The SIFT matching algorithm usually adopts Euclidean distance as the similarity measurement of the key point. If the ratio of two key points with the closest Euclidean distance to the candidate key point for matching is less than a certain threshold (always taken as 0.9–1.0), then one of these two key points is considered a successful match. The number of matching points can be reduced by lowering the threshold ratio, thus stabilizing the SIFT matching process. In the proposed approach, the matching between RTI and GRID is simplified as the calculation of the Euclidean distance in the set of key points between RTI and GRID. A best-bin-first matching search algorithm with a 128-dimensional feature vector is presented as well, and all these strategies reduce the duration of searching and matching.

The correlation calculation based on Euclidean distance exhibits gross error. Handling mismatch is another important process during SIFT matching; normally, the SIFT matching algorithm follows a robust random sample consensus (RANSAC) [37,38] algorithm to address the gross error in SIFT matching.

The image searching and matching processes intersect, as depicted in Figure 5. We can determine the match status according to the number of the matched key points between RTI and GRID. The number of matched points can be used to guide the search direction based on the LRS in a large file. Figure 6 presents an example of the number variance of the matched points. A total of 59 images from GRID are considered candidate images to match RTI.

The number of matched points increases when the GRID image is close to RTI. The entire matching process uses the same parameter of the SIFT algorithm and RANSAC. Thus, the matched image has the highest number of matching points. The number of matched points increases rapidly when the candidate image is in view of RTI; this step also helps determine the direction of image searching. The searching and matching time in all 59 images is 0.43 s; in real-time processing, these procedures stop after the image either matches well or fails. The searching and matching time is shorter than that of the SIFT matching of all the candidate images. The 59 GRID images cover a distance of approximately 300 m, which can be considered the largest error in the GPS/IMU position serving as the initial input for image interval retrieval.

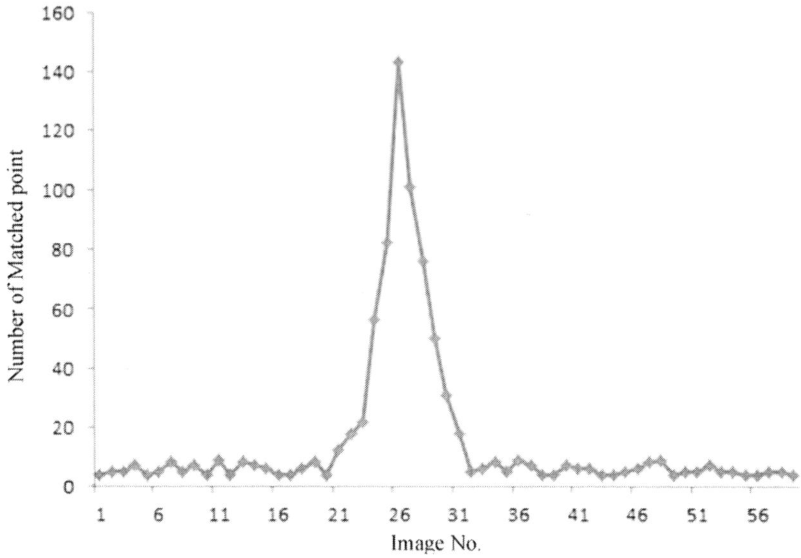

Figure 6. Fast search and large file location based on spatial query.

2.4. 3D Navigation Calculation: From GRID to RTI

In the proposed imaging sensor-aided vision navigation system, a camera is fixed to the body of a moving vehicle, as described by a right-handed coordinate system. We define five coordinate systems to derive the navigation parameters from GRID to RTI, as shown in Figure 7.

GP(XG,YG,ZG) is the Cartesian coordinate of the geodetic coordinate system(GCS). GO is the coordinate origin and its coordinates in GCS are PG, which is adopted by the GRID and the moving platform. VP(xV,yV,zV) is the body coordinate system(BCS) of a moving platform where the coordinate origin VO is defined as the intersection of two perpendicular lines of the cut planes through the center of gravity and the body; its coordinates in BCS are PV. The XV axis is parallel to the longitudinal axis of the body, which is the forward direction of the vehicle. The YV axis is perpendicular to the longitudinal axis of the vehicle body and parallel to the cutting plane, which is located to the right of the vehicle. The ZV axis is downward perpendicular to the cut plane. CP(xC,yC,zC) is the camera coordinate system(CCS), where coordinate origin CO is the projection center of the camera, and its coordinates in CCS are PC. The XC axis is parallel to the direction of the camera scan lines and to the direction of the increasing scan pixels. The YC axis is perpendicular to the direction of the camera scan lines and to the direction of the increasing scan lines. The ZC axis is perpendicular to the silicon target plane and to the gaze direction of camera imaging system. IP(xI,yI) is the image coordinate system (ICS) that defines the location coordinates of the optical imaging plane of the internal camera image, where coordinate origin IO is the focus on the image plane of the imaging system; its coordinates in ICS are PI. The XI axis is parallel to the direction of the camera scan lines and to the direction of the increasing scan pixels. The YI axis is perpendicular to the direction of the camera scan lines and to the direction of the increasing scan lines. BP(c,r) is the pixel coordinate system(PCS) that defines the pixel coordinate, where coordinate origin BO is located in the top-left corner of the image; its coordinates in PCS are PB. The c axis is parallel to the direction of the camera scan lines and to the direction of the increasing scan pixels. The r axis is perpendicular to the direction of the camera scan lines and to the direction of the increasing scan lines. Figure 7 indicates that ψ, θ, φ, VXC, VYC, and VZC are considered to define the installation status of imaging cameras in the camera coordinate system.

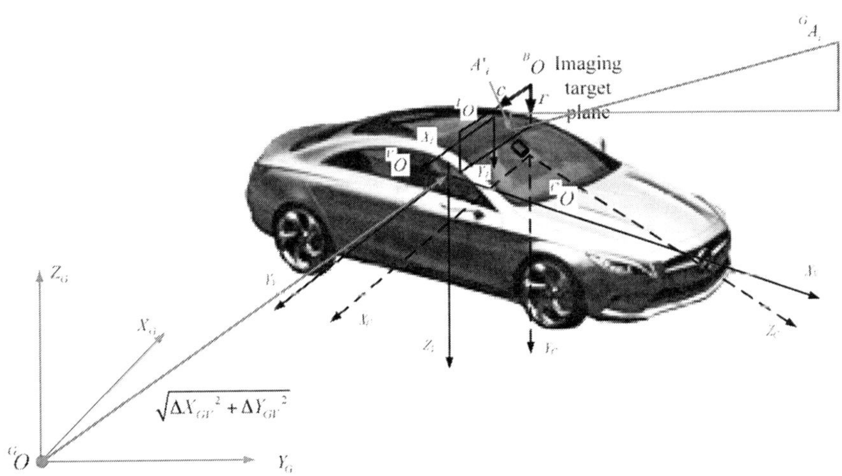

Figure 7. Definition of a coordinate system.

The following paragraphs describe a few of the main transformations of the aforementioned coordinate system.

(1) *From the body coordinate system into the camera coordinate system*: the body coordinate system can be transformed into the pixel coordinate system. The transition from the former into the camera coordinate system includes translation and rotation, as indicated in Equation (1)

$$^{C}P_{(x_C,y_C,z_C)} = {}^{C}_{V}R(^{V}P_{(x_V,y_V,z_V)} - {}^{V}P_C) \tag{1}$$

where VPC is the coordinates of CO in the body coordinate system. CVR is the rotation matrix, as per Equation (2)

$${}^{C}_{V}R = R_Y(\psi)R_X(\theta)R_Z(\phi) \tag{2}$$

(2) *From the camera coordinate system into the image coordinate system*: this transformation is based on the principle of perspective projection, where the object distance is far greater than the focal length. Thus, the pinhole imaging model can replace the perspective projection model. The image coordinate can be calculated with Equation (3)

$$\begin{cases} x_I = f\dfrac{x_c}{z_c} \\ y_I = f\dfrac{y_c}{z_c} \end{cases} \tag{3}$$

(3) *From the image coordinate system into the pixel coordinate system*: the pixel coordinate system can be calculated from the image coordinate system through a 2D scale and translation, as indicated in Equation (4)

$$^{B}P = \begin{bmatrix} N_r & 0 \\ 0 & N_c \end{bmatrix} \begin{bmatrix} x_I \\ y_I \end{bmatrix} + {}^{B}P_I \tag{4}$$

where Nr and Nc are the numbers of rows and columns of width and length in the image coordinate system, respectively.BPI=[BCI,BRI]T is the offset of the main point position in the column and row directions.

In the transformation from Equations (1) to (4) explained above, ψ, θ, φ, VXC, VYC, and VZC are usually called the outer orientation elements of an imaging camera. f, Nr, Nc, BCI, and BRI are known as the inner orientation elements of such a camera.

(4) From the body coordinate system into the geodetic Cartesian coordinate system: the matched image is geo-referenced in the geodetic Cartesian coordinate system. All the matched key points have Cartesian coordinates GAi(XGAi,YGAi,ZGAi)that are transferred to the corresponding points in RTI. Thus, we can convert GAi to VAi in the body coordinate system of the moving platform with Equation (5):

$$\begin{cases} X_{GV} = X_V \cos\alpha + \Delta X_G \\ Y_{GV} = Y_V \cos\alpha + \Delta Y_G \\ Z_{GV} = -Z_V \end{cases} \quad (5)$$

where ΔXGV and ΔYGV are unknown quantities, and α is the heading of the moving platform. Then, we can convert VAito CAi in the camera coordinate based on Equations (1) and (2). We can also transform CAi into IAi in the image coordinate system and BAi in the pixel coordinate system on the basis of Equations (3) and (4). We finally complete the transformation of the key points CAi(xCAi,yCAi,zCAi) in the camera coordinate system to BAi(xBAi,yBAi) in the pixel coordinate system, as shown in Equation (6):

$$\begin{cases} x_{BAi} = f\dfrac{x_{CAi}}{z_{CAi}} N_r + {}^B C_I \\ y_{BAi} = f\dfrac{y_{CAi}}{z_C} N_c + {}^B R_I \end{cases} \quad (6)$$

The BA'i(xBAi,yBAi) in the pixel coordinate system of RTI is determined through SIFT matching, and the corresponding GAi(XGAi,YGAi,ZGAi) in the geodetic Cartesian coordinate system is also identified from the GRID. Both outer orientation elements (ψ, θ, φ, VXC, VYC, and VZC) and inner orientation elements (f, Nr, Nc, BCI, and BRI) can be effectively calibrated in a high-precision calibration site. Each pair of matched key points can generate two equations based on Equation (6), and we can build the equations with all the matched key points with unknown elements ΔXGV, ΔYGV, and α. Thus, we can employ least squares' adjustment to solve these unknown elements, which are the navigation parameters of the moving platform in the geodetic Cartesian coordinate system. To obtain a robust result, at least five key points should be used in the solving process.

3. EXPERIMENTAL SECTION AND DISCUSSION

The proposed imaging sensor-aided vision navigation approach has two key performance indicators that affect its actual application. One is the efficiency of the image searching and matching process, which is related to the GRID data organization model, and the other is navigation accuracy. We design two experiments to assess the proposed approach in terms of efficiency and accuracy. An L-MMS with six cameras is developed to collect GRID data; the GRID accuracies are 0.5 m in position and 0.3° in orientation. The GPS/IMU utilized in the L-MMS is SPAN® IMU-FSAS™. The L-MMS is also used to verify the accuracy of the proposed vision navigation method based on GRID, whereas the high-precision GPS/IMU navigation data are considered the truth. One camera is simulated as the real-time imaging sensor.

3.1. GRID Efficiency Experiment

The GRID data organization method directly affects the efficiency of searching and matching between RTI and GRID. Thus, two experiments were designed to validate the proposed method. The computer is configured with a Pentium 4 2.0 GHz CPU, 1 GB memory, and a direct fiber 80 TB disk array with Windows XP Professional. We take three sets of GRID with data volumes that range from 1 GB and 100 GB to 1 TB by utilizing the proposed data organization model. This model represents the road, street (town), and county levels, as shown in Table 1.

Table 1. Experimental GRID data set.

Data Set	Data Volume (G)	Mileage (km)	Data Level
I	1.9	1.8	road
II	120.0	115.3	street (town)
III	1035.0	1027.7	county

3.1.1. Comparison of the DS and Quad-Tree Indices for Image Spatial Query

The DS index converts the spatial query in 2D to a 1D space. Table 2 indicates the query response times after clicking the 2D navigation map as an image query input.

Table 2. Comparison of image spatial query times of the DS tree and the quad-tree.

Data Set	Index Model	Query Time (s)
I	Quad-tree	0.50
	DS tree	0.21
II	Quad-tree	0.52
	DS tree	0.22
III	Quad-tree	0.55
	DS tree	0.24

Image spatial query, which is based on the DS index, is faster than the quad-tree index. Furthermore, the proposed GRID data organization model does not differ significantly in terms of data volume because of the large file model. Each query is restricted to a specific large file.

3.1.2. Comparison of Image Retrieval Efficiencies in Large and Small GRID File Storage Models

Image retrieval includes the image searching and image matching processes. Once the current searching image from GRID is matched to RTI, image searching ceases. We compare the image retrieval efficiency of the proposed large file model with that of the single-image file model. The average retrieval times of the three data sets are listed in Table 3.

Table 3. Comparison of image retrieval times in the large file model and the single-image file model.

Data Set	Index Model	Image Retrieval Time (s)
I	Single-image file	1.25
	Large file	0.78
II	Single-image file	4.86
	Large file	0.82
III	Single-image file	18.24
	Large file	0.86

The proposed large file model based on road sections is obviously advantageous over the single-image file model regardless of whether the files are stored in a database system or in a file directory. The image retrieval time of a large file model-based GRID is less than 1 s given three data sets. The retrieval time of the single-image file-based GRID increases with the data volume. The efficiency of the proposed GRID data organization model can meet the requirement of vision navigation based on this database.

3.2. Accuracy Analysis of GRID-Based Vision Navigation

To assess the accuracy of the proposed vision navigation method, we use L-MMS to collect data on a single block twice at different times. One dataset is utilized to build the GRID as the proposed approach. Another dataset is employed to simulate RTI. A total of 300 images are regarded as RTIs for searching and matching with GRID. The image interval distance is set as 5 m, and the total time for 300 images is approximately 300 s within a distance of 1500 m. Then, the navigation parameters of the moving platform are calculated from the 300 images with the use of the proposed approach. The navigation results are compared with the tightly coupled integration of GPS/IMU via Inertial Explorer® software. Figure 8 presents the absolute value of the relative error in *x*, *y*, and *H*.

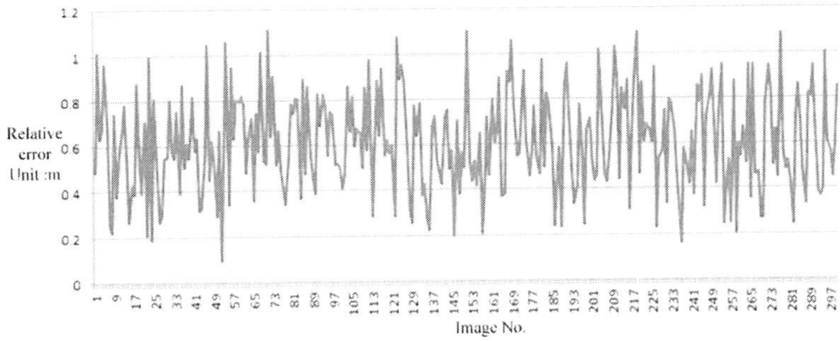

(a)

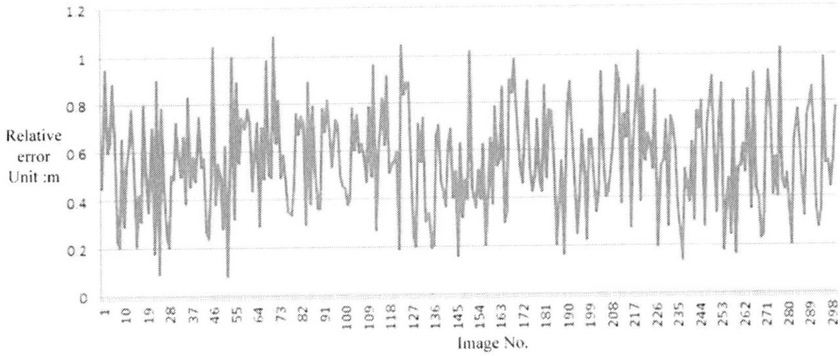

(b)

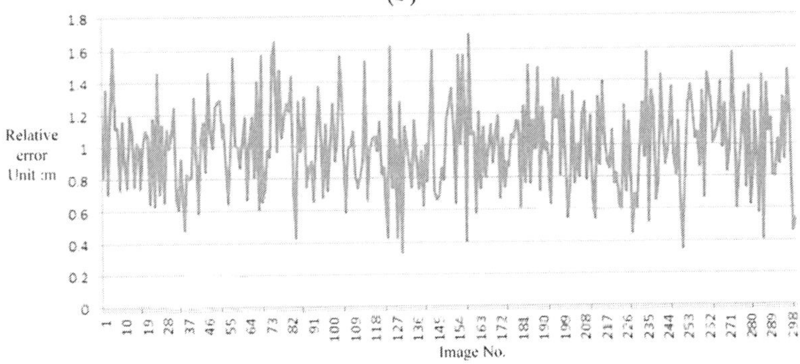

(c)

Figure 8. Accuracy of the proposed vision navigation approach: (**a**) *x* direction; (**b**) *y*direction; (**c**) *H*.

The accuracy of the proposed imaging sensor-aided vision navigation method is less than 1.2 m in the x and y directions.*H* is less than 1.8 m. This accuracy is approximately double the accuracy of GRID because matching errors are detected in the key points between RTI and GRID.

3.3. Discussion

The use of a map as a sensor is a new ADAS trend. GRID is a map database that contains highly accurate position data and real environment information [39]; it can be used as a high-precision map database for ADAS. Although we have presented a basic framework of vision navigation with GRID, some challenges are still encountered for widely commercial applications in the navigation system. Nonetheless, the GRID data volume is extensive, as discussed in this paper; that of one county reaches the TB level. The current navigation database with a vector map always covers one county. Storing the high volume of GRID data on the navigation platform is a significant problem. The fast development of wireless Internet, such as 4G LTE and 5G LTE, is a way to realize online GRID services for vision navigation [40].

Although the GRID can provide real environment information for vision navigation, changes in road scenes are another challenge in imaging sensor-aided navigation. It is common that one road can change completely in a developing country. Such an altered scene complicates image retrieval and induces failure in certain cases. Rapid data collection and updating are important in GRID maintenance. The incremental update technology for a navigation map can also be used for this database. However, determining scene changes is particularly difficult in data collection and updating. Volunteered geographic information can harness tools to create, assemble, and disseminate geographic data provided voluntarily by individuals [41]; this information can serve as a crowdsourcing mode to determine scene changes for updating.

In fact, no error accumulates in the proposed technology. The reliability of localization after long periods of time depends on the image matching algorithms. Image matching is also a critical issue in GRID-based vision navigation [42]. Although we have introduced SIFT matching for RTI and GRID, challenges are still encountered with regard to matching speed and reliability, particularly given a wide image searching range. Such difficulties are experienced in vision navigation with low-cost GPS/IMU. Poor position inputs widen the range of image searching. Such a range worsens the reliability of image matching. The improved SIFT algorithms with multicore parallel computing [43,44] can be discussed further to improve the efficiency and reliability of this matching.

GRID-based vision navigation can be integrated with GPS/IMU, where the GRID can serve as a new sensor for federal Kalman filtering (FKF) in navigation calculation [45]. Our paper merely discusses vision navigation during GPS loss while one or more GPS satellites remain. Thus, we can further combine the result of vision navigation and of GPS/IMU based on FKF. The fusion navigation result can serve as vision navigation feedback to help narrow the searching range. The reliability and robustness of fusion navigation can also be enhanced with increased accuracy.

4. CONCLUSIONS

GRIDs are expanding worldwide with high resolution and accuracy, included Google Street View [46] and the Tencent Panoramic Image Service. The rapid development in global geo-referenced databases enables vision navigation with GRID. In this paper, we proposed a vision navigation approach based on this database to facilitate continuous and robust navigation for GPS and IMU sensors under harsh environmental conditions. The framework and navigation calculating model of GRID-aided vision navigation is established with sequence images from land-based multiple sensor-integrated mobile mapping systems. A highly efficient GRID storage management model is also developed based on the linear index of a road segment; large files are created to achieve fast image search and retrieval. This model can perform image searching and matching in 1 s regardless of data volume. A calculation method for 3D navigation parameters from GRID to RTI is also presented for imaging sensor-aided vision navigation with matched GRID images. The final result of vision navigation based on a GRID with 0.5 m accuracy indicates a 1.2 m accuracy in plane and 1.8 m accuracy in height under GPS loss in 5 min and within 1500 m. The proposed approach can be used to improve navigation reliability under harsh environmental conditions where GPS and IMU are ineffective, such as ADAS. When integrated with GPS/IMU, this method can also be employed to enhance navigation accuracy for L-MMS.

ACKNOWLEDGMENTS

This work is supported by the National Natural Science Foundation of China (No. 41271452 and No. 41271431).

AUTHOR CONTRIBUTIONS

Yan Li wrote the main program and most of the paper; Qingwu Hu conceived the study and designed the experiments; Meng Wu studied the image matching and 3D Navigation calculation approach. Yang Gao was responsible for the part of data organization model.

CONFLICTS OF INTEREST

The authors declare no conflict of interest.

REFERENCES

1. Duong, T.T.; Chiang, K.W. Real-time INS/GPS Integration with Optimal On-line Smoothing for Mobile Mapping Systems Utilizing a Low Cost MEMS IMU. *J. Aeronaut. Astronaut. Aviat. Ser. A* **2013**, *45*, 187–193.

2. Li, D.; Hu, Q.; Guo, S.; Chen, Z. Land-based mobile mapping system with its applications for the Olympic Games.*Chin. Sci. Bull.* **2009**, *54*, 2286–2293.

3. Zhu, Z.S.; Su, A.; Liu, H.B.; Shang, Y.; Yu, Q. Vision navigation for aircrafts based on 3D reconstruction from real-time image sequences. *Sci. China Technol. Sci.* **2015**, *58*, 1196–1208.

4. Guerrero, L.A.; Vasquez, F.; Ochoa, S.F. An indoor navigation system for the visually impaired. *Sensors* **2012**, *12*, 8236–8258.

5. Cooper, A.J.; Redman, C.A.; Stoneham, D.M.; Gonzalez, L.F.; Etse, V.K. A dynamic navigation model for unmanned aircraft systems and an application to autonomous front-on environmental sensing and photography using low-cost sensor systems. *Sensors* **2015**, *15*, 21537–21553.

6. Godha, S. *Performance Evaluation of Low Cost MEMS-Based IMU Integrated with GPS for Land Vehicle Navigation Application*; UCGE Report; Department of Geomatics Engineering: Calgary, AB, Canada, 2006.

7. DeSouza, G.N.; Kak, A.C. Vision for mobile robot navigation: A survey. *IEEE Trans. Pattern Anal. Mach. Intell.* **2002**, *24*, 237–267.

8. Bonin-Font, F.; Ortiz, A.; Oliver, G. Visual navigation for mobile robots: A survey. *J. Intell. Robot. Syst.* **2008**, *53*, 263–296.

9. Güzel, M.S. Autonomous vehicle navigation using vision and mapless strategies: a survey. *Adv. Mech. Eng.* **2013**, *5*, 234747.

10. Tang, J.; Chen, Y.; Chen, L.; Liu, J.; Hyyppä, J.; Kukko, A.; Kaartinen, H.; Hyyppä, H.; Chen, R. Fast Fingerprint Database Maintenance for Indoor Positioning Based on UGV SLAM. *Sensors* **2015**, *15*, 5311–5330.

11. Guerra, E.; Munguia, R.; Grau, A. Monocular SLAM for autonomous robots with enhanced features initialization.*Sensors* **2014**, *14*, 6317–6337.

12. Mendoza, G.C.; Davison, A. *Robot SLAM and Navigation with Multi-camera Computer Vision*; Imperial College London (University of London): London, UK, 2012.

13. Van Pham, B.; Lacroix, S.; Devy, M. Vision-based absolute navigation for descent and landing. *J. Field Robot.* **2012**, *29*, 627–647.

14. Rau, J.Y.; Habib, A.F.; Kersting, A.P.; Chiang, K.-W; Bang, K.; Tseng, Y.; Li, Y. Direct sensor orientation of a land-based mobile mapping system. *Sensors* **2011**, *11*, 7243–7261.

15. Lillo A, D.; Daptardar, A.; Thomas, K.; Thomas, K.; Storer, J.A.; Motta, G. Compression-based tools for navigation with an image database. *Algorithms* **2012**, *5*, 1–17.

16. Choi, K.; Tanathong, S.; Kim, H.; Lee, I. *Realtime Image Matching for Vision Based Car Navigation with Built-in Sensory Data*; ISPRS Annals of the Photogrammetry, Remote Sensing and Spatial Information Sciences: Antalya, Turkey, 2013.

17. Larsson, A.; Varhelyi, A. *Long-Term Effects of Advanced Driver Assistance Systems–Literature Review*; SP7 Working Report of the InteractIVe Project; InteractIVe Consortium: Lund, Sweden, 2012.

18. Daza, I.G.; Bergasa, L.M.; Bronte, S.; Yebes, J.J.; Almazán, J.; Arroyo, R. Fusion of optimized indicators from Advanced Driver Assistance

Systems (ADAS) for driver drowsiness detection. *Sensors* **2014**, *14*, 1106–1131.

19. Bhatnagar, G.; Liu, Z. A novel image fusion framework for night-vision navigation and surveillance. *Signal Image Video Process.* **2015**, *9*, 165–175.

20. Li, X.; Wang, J. Image matching techniques for vision-based indoor navigation systems: A 3D map-based approach1. *J. Locat. Based Serv.* **2014**, *8*, 3–17.

21. Yu, Y.; Huang, K.; Chen, W.; Tan, T. A novel algorithm for view and illumination invariant image matching. *IEEE Trans. Image Process.* **2012**, *21*, 229–240.

22. Leutenegger, S.; Chli, M.; Siegwart, R.Y. BRISK: Binary robust invariant scalable keypoints. In Proceedings of the 2011 IEEE International Conference on Computer Vision (ICCV), Barcelona, Spain, 6–13 November 2011; pp. 2548–2555.

23. Puente, I.; González-Jorge, H.; Martínez-Sánchez, J.; Arias, P. Review of mobile mapping and surveying technologies.*Measurement* **2013**, *46*, 2127–2145.

24. Ou, J.; Qiao, G.; Bao, F.; Wang, W.; Di, K.; Li, R. A new method for automatic large scale map updating using mobile mapping imagery. *Photogramm. Rec.* **2013**, *28*, 240–260.

25. Mao, Q.; Zhang, L.; Li, Q.; Hu, Q.; Yu, J.; Feng, S.; Ochieng, W.; Gong, H. A Least Squares Collocation Method for Accuracy Improvement of Mobile LiDAR Systems. *Remote Sens.* **2015**, *7*, 7402–7424.

26. Li, D.R.; Hu, Q.W. DMI Based Spatial Information Service. Geomat. *Inf. Sci. Wuhan Univ.* **2007**, *32*, 377–380.

27. Nagaki, K. *Evolution of In-Car Navigation Systems*; Handbook of Intelligent Vehicles; Springer London: London, UK, 2012; pp. 463–487.

28. Xu, J.H; Li, Q.Q.; Song, Y.; Zhen, N.B. Kiwi Data Format Based Map Display Research. *Geomat. Inf. Sci. Wuhan Univ.***2005**, *30*, 866–868.

29. Graettinger, A.; Morrison, A.; Parker, S.; Nianbo, Z. Translating Transportation Data between Linear Referencing Systems of Dissimilar

Resolution. *Transp. Res. Rec. J. Transp. Res. Board* **2013**, *2399*, 103–111.

30. Ryals, Z.T. *A Technique for Merging State and Non-State Linear Referencing Systems*; The University of Alabama Tuscaloosa: Tuscaloosa, AL, USA, 2011.

31. Graettinger, A.; Qin, X.; Spear, G.; Parker, S.; Forde, S. Combining state route and local road linear referencing system information. *Transp. Res. Rec. J. Transp. Res. Board* **2009**, *2121*, 152–159.

32. Bigham, J.; Kang, S. *Building a Highway Linear Referencing System from Â Preexisting Reference Marker Measurements for Â Transportation Data Management*; Institute of Transportation Studies, UC Berkeley: Berkeley, CA, USA, 2013.

33. Lowe, D.G. Object recognition from local scale-invariant features. In Proceedings of the Seventh IEEE International Conference on Computer Vision, Kerkyra, Greece, 20–27 September 1999; Volume 2, pp. 1150–1157.

34. Lowe, D.G. Local feature view clustering for 3D object recognition. In Proceedings of the 2001 IEEE Computer Society Conference on Computer Vision and Pattern Recognition, Hawaii, HI, USA, 8–14 December 2001; Volume 1, pp. I-682–I-688.

35. Huo, J.; Chen, P.; Zhang, P.; Yang, L. A review of improved SIFT algorithm in real-time matching. *Inf. Sci. Manag. Eng.* **2014**, *62*.

36. Zhu, Q.; Liu, C.; Cai, C. A Novel Robot Visual Homing Method Based on SIFT Features. *Sensors* **2015**, *15*, 26063–26084.

37. Gao, H.; Xie, J.; Hu, Y.; Yang, Z. Hough-RANSAC: A Fast and Robust Method for Rejecting Mismatches. In *Pattern Recognition*; Springer Berlin Heidelberg: Changsha, China; 17–19; November; 2014; pp. 363–370.

38. Chum, O.; Matas, J.; Kittler, J. Locally Optimized RANSAC. In *Pattern Recognition*; Springer Berlin Heidelberg: Magdeburg, Germany; 10–12; September; 2003; pp. 236–243.

39. Li, K.; Tan, H.S.; Misener, J.A.; Hedrick, J.K. Digital map as a virtual sensor–dynamic road curve reconstruction for a curve speed assistant. *Veh. Syst. Dyn.* **2008**, *46*, 1141–1158.

40. Sulyman, A.I.; Nassar, A.M.T.; Samimi, M.K.; Maccartney, G.R. Radio propagation path loss models for 5G cellular networks in the 28 GHz and 38 GHz millimeter-wave bands. *IEEE Commun. Mag.* **2014**, *52*, 78–86. [

41. Goodchild, M.F. Citizens as sensors: The world of volunteered geography. *GeoJournal* **2007**, *69*, 211–221.

42. Valgren, C.; Lilienthal, A. Sift, Surf, and Seasons: Long-term outdoor localization using local features. In Proceedings of the 3rd European Conference on Mobile Robots, Freiburg, Germany, 2007.

43. Cao, H.; Chen, J. Multicore computing for SIFT Algorithm in MATLAB® Parallel Environment. In Proceedings of the 2012 IEEE 18th International Conference on Parallel and Distributed Systems (ICPADS), Singapore, 17–19 December 2012; pp. 924–929.

44. Qasaimeh, M.; Sagahyroon, A.; Shanableh, T. A parallel hardware architecture for Scale Invariant Feature Transform (SIFT). In Proceedings of the 2014 International Conference on Multimedia Computing and Systems (ICMCS), Marrakech, Morocco, 14–16 April 2014; pp. 295–300.

45. Yuan, G.; Yuan, K.; Zhang, H. A Variable Proportion Adaptive Federal Kalman Filter for INS/ESGM/GPS/DVL Integrated Navigation System. In Proceedings of the 2011 Fourth International Joint Conference on Computational Sciences and Optimization (CSO), Yunnan, China, 15–19 April 2011; pp. 978–981.

46. Anguelov, D.; Dulong, C.; Filip, D.; Frueh, C.; Lafon, S.; Lyon, R.; Ogale, A.; Vincent, L.; Weaver, J. Google street view: Capturing the world at street level. *Computer* **2010**, *43*, 32–38.

Chapter 8

Vision-Based Sensor for Early Detection of Periodical Defects in Web Materials

Francisco G. Bulnes *, Rubén Usamentiaga, Daniel F. García and Julio Molleda

Department of Computer Science, University of Oviedo, Campus de Viesques, Gijón 33204, Spain

ABSTRACT

During the production of web materials such as plastic, textiles or metal, where there are rolls involved in the production process, periodically generated defects may occur. If one of these rolls has some kind of flaw, it can generate a defect on the material surface each time it completes a full turn. This can cause the generation of a large number of surface defects, greatly degrading the product quality. For this reason, it is necessary to have a system that can detect these situations as soon as possible. This paper presents a vision-based sensor for the early detection of this kind of defects. It can be adapted to be used in the inspection of any web material, even when the input data are very noisy. To assess its performance, the sensor system was used to detect periodical defects in hot steel strips. A total of 36 strips produced in ArcelorMittal Avilés factory were used for this purpose, 18 to determine the optimal configuration of the proposed sensor using a full-factorial experimental design and the other 18 to verify the validity of the results. Next, they were compared with those provided by a commercial system used worldwide, showing a clear improvement.

KEYWORDS

vision sensors; intelligent systems; automated defect detection; pattern recognition

1. INTRODUCTION

Quality control in industrial environments is an essential aspect, both to improve product quality and to reduce costs caused by discarded defective products. The current trend is to replace human experts for automated systems, since they are cheaper, take less time and can work even in the most dangerous environments [1]. This study focuses on quality control during production of web materials using a computer vision system. Such materials can be defined as those which are produced by continuous rolling processes, such as paper, plastic, fabric or metal. The rolls that are in direct contact with the web material to be produced can be the source of serious defects. When their surfaces have some kind of flaw, they can generate a surface defect on the rolled product each time they complete a full turn, generating a periodical pattern as shown in Figure 1. In these cases, this situation is repeated until the defective roll is replaced. For this reason, it is very important to make early detection of such defects during the production of web materials. The sooner this situation is detected, the sooner the defective roll may be replaced, greatly reducing the productivity loss.

There are several studies about the development of vision-based sensor systems for industrial use, such as inspection tasks [2] or defect detection tasks in many different fields, like satin glass [3], welds [4], bearings [5] or ship hulls [6]. Depending on each case, the design of the vision system (cameras, lenses, lighting, *etc.*) is different. It is critical to build the most suitable vision system in each case in order to obtain satisfactory results [7]. In the particular case studied here, the sensing tasks are carried out using several cameras, which take images of both surfaces of the material to be inspected. The areas inspected by the cameras are overlapped in order to ensure complete coverage, as shown in Figure 2. These images are used by a software system to perform other typical tasks of these systems, namely fault detection, characterization of defects, feature extraction and classification [8]. Using an inspection system that performs these tasks, the technicians responsible for quality control can obtain information about the position and class (or type) of each defect present in the produced material. Some of these systems are able to determine whether a surface defect is a roll mark or not, based on its features. Thus, a technician can find out that at least one roll is defective. In production lines in which only one roll can generate defects, this information may be sufficient. However, in more complicated cases, where several rolls can be the source of such defects, more information about the periodical defects generated is required to determine which roll must be replaced.

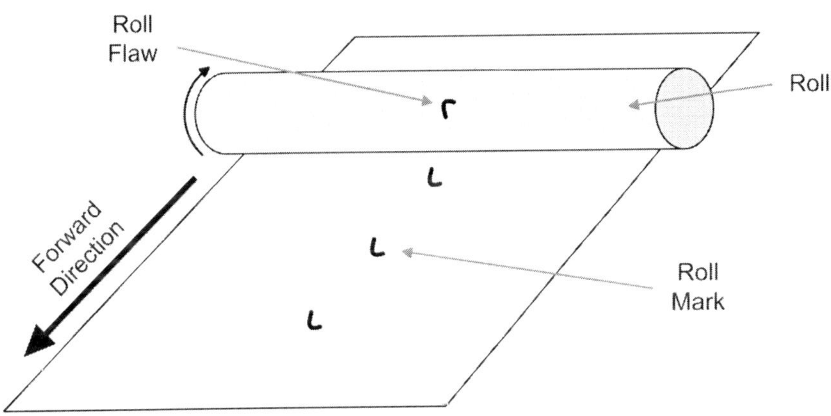

Figure 1. Defect generation due to a defective roll.

In the most adverse situations, there may be several defective rolls generating defects simultaneously. To handle these situations, it is desirable to cluster the defects that have been generated by the same flaw in the same roll. The characteristics of each cluster of defects can be used to identify the defective roll. The separation between defects of the same cluster may be used for this. Normally this separation coincides with the perimeter of the roll that generated the periodical defect (if the radius of each roll involved in the production process is different, it would identify the defective roll unequivocally).

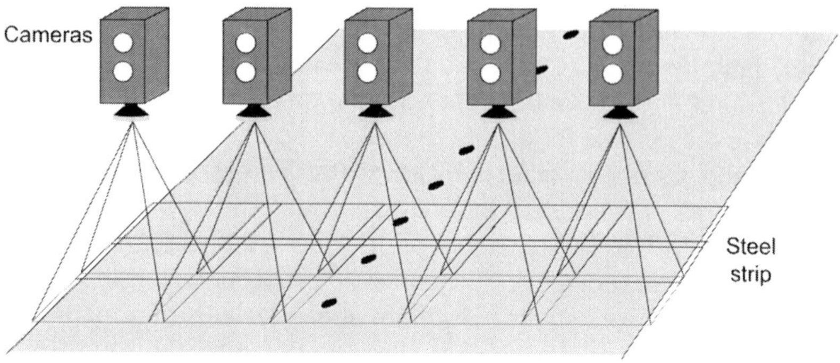

Figure 2. Cameras used in top surface inspection.

An additional phase for the design of a sensor capable of detecting periodical defects is proposed: the clustering phase. Its objective is the detection of surface defects generated periodically based on data provided by traditional inspection systems: position, type and features of each surface defect. This latter phase is performed by a software module that executes a clustering method. The main problem to develop this method is that the input data may be very different depending on the web material being inspected. As the inspection system and the production process may differ, the characteristics of the input data can vary greatly. The number of surface defects per square meter and the types in which they can be classified may be very different. In addition, inspection systems always add error to their measurements, which are different depending on the system used [9]. Therefore, the clustering method must take this into account. It should be able to be adapted to work with input data with different noise levels. Furthermore, in order to calculate the results quickly, consumption of computational resources and memory must be low. Thus, early detection of periodical defects can be achieved even without using a powerful computer.

A method that can detect such defects quickly even in noisy environments is proposed. In order to assess the proper operation of the whole sensor, it has been used in a real situation: the detection of periodical defects in hot steel strips. In this case, there are several rolls which can generate periodical marks (in our example there are 7 rolls) and the input data are very noisy. For these reasons, each strip can have a very high number of surface defects, making the task of detecting periodical defects much more difficult.

Less evolved versions of this method have been published previously by the authors (Bulnes *et al.* [10], Bulnes *et al.* [11], Bulnes *et al.* [12]). With more features to define the periodical defects (Subsection 3.1), improvements in the clustering algorithm (Section 3.2) and more configuration parameters (Section 3.3), better outputs are achieved. In addition, all steps taken by the proposed method are explained in detail, in order that anyone can reproduce it.

This paper is organized as follows. In Section 2, we review previous work related to the detection of periodical defects in web materials. The proposed method to perform periodical defect detection is introduced in Section 3. Specifically, it describes the way the information is managed, the detection algorithm and its parametric optimization. Section 4 presents the utilization of the proposed method to a practical application: hot steel strips, in order to verify that it can be used in a real industrial environment. Finally, in Section 5, we summarize the main conclusions of the paper.

2. RELATED WORK

The available literature about the detection of periodical defects is scarce. Although many of the commercial systems that are used to carry out inspection tasks of web materials are able to detect these defects, the information about this process is not public. However, there are some published studies where periodical defects are detected in different types of web materials, which are discussed below.

2.1. Paper

Paper is usually made using wood pulp. A typical paper machine consists of the wet end (or forming section), the press section and the drier [13]. In the wet end, the sheet of paper is formed by passing the mixture of pulp and water between two porous belts called forming fabrics. In the press section, most of the water is removed by applying pressure to the sheet using several rollers. Finally, any remaining moisture is evaporated using heated rolls in the drier. Commonly, a final stage is also used: the calender section, where the paper can obtain a smoother and more uniform surface by applying pressure to the already dry paper using rolls.

Although many rolls are involved in the paper making process, most of the periodical defects are produced in the press section, because a high pressure is applied when the paper is still in a delicate and deformable state [14]. The authors of the said paper were able to detect the periodical defects using two-dimensional fast Fourier transforms. However, several periodical and pseudo-periodical defects may appear in the wet end too. In [15,16], periodical defects generated in the wet end were detected using Fourier analysis, and periodical defects generated in the wet end and press section were also detected using the same technique by Chaplin [17].

2.2. Textiles

Depending on the type of fabric, the production process can vary greatly. Moreover, unlike what happens in the case of paper or in the case of hot metal, textiles are not in such a delicate state during the production process, which reduces the possibility of generation of periodical defects. During the production of textiles, inspection systems detect and classify surface defects [18], but rarely is the task of detecting periodical defects performed.

The first study that includes the detection of periodical defects was performed by Toba [19]. In this study, autocorrelation functions were used to inspect products and carry out the detection of periodical defects. This method was also used by Drean *et al.* [20] to detect periodic irregularities in textile surfaces. Furthermore, a variance analysis was also used. Another recent example of the defects that were made periodically is studied by Kianiha *et al.* [21], where fabric products are analyzed using image analysis methods.

2.3. Metal

During the production of metal products, such as rods, tubes or strips, the metal leaves the furnace at very high temperatures. While still hot, using a series of rolls, its desired shape is obtained. At this point the metal is very malleable, which favors the appearance of periodical defects due to flaws in any of the rolls. In these cases, the inspection task is usually performed using computer vision techniques, although it depends ultimately on the type of product to be inspected.

The metal product for which most studies of this kind have been made are steel or iron strips. In Duan *et al.* [22], a neural network is used to classify each surface defect within a strip depending on its characteristics. Since all defects generated periodically by defective rolls have similar characteristics, this classification can be used to warn about that situation. Similarly periodical defects are detected in [23]. However, Shigeno *et al.* [24] proposes a method to detect periodical defects based on the position of each surface defect, instead of using its characteristics. This method is similar to the one proposed in this paper, described in Section 3.

This type of detection is also performed during the production of elongated metal products such as tubes or wires. In Traxler *et al.* [25] a method for the detection of defective rolls in the production of seamless steel tubes is proposed. In the said study, the detection is performed taking into account the characteristics of each surface defect and also its position. Moreover, Park *et al.* [26] proposes two methods for detecting and identifying periodical defects in steel wire rods. Their detection is performed using a Discrete Wavelet Transform, and their identification is carried out using a Discrete Fourier Transform.

2.4. Other

Periodical defects can also occur in other types of web materials. For example, in Oerley *et al.* [27] periodical defect detection in plastic film material is performed using a correlation function.

In Caron *et al.* [28], the inspection of preweathered zinc is presented. The inspection system used in this case detects periodical defects within zinc strips which are submitted to a process (called preweathering) that gives the strip a slate gray color. During both the production process of the strip and preweathering, periodical defects may appear. In this study, the surface defects are detected using computer vision and classified based on their characteristics. The system only warns about defective rolls when there are defects which have been classified as “roll mark”.

3. PROPOSED APPROACH

The proposed approach will perform the detection of periodical defects using a method whose objective is to look for surface defects that meet certain conditions. In this section the main contributions of the paper are presented: the way to represent a periodical defect as a pattern defined by 9 features and the parameterizable backtracking based algorithm responsible for carrying out the detection and characterization of these defects.

As mentioned above, the uncertainty introduced by the inspection system must be taken into account. Figure 3 shows four different individual defects generated by a defective roll during the production of a steel strip. They all have the same shape, but the dimensions estimated for each of them by the inspection system are very different. Consequently, the position assigned to each is not correct, since it is calculated as the central point of the square surrounding the defect. Although these cases are extreme, the inspection system always introduces an uncertainty in the characteristics of each individual defect (this error may be higher or lower depending on the web material inspected and the inspection system used). Since the type assigned to each surface defect depends on its characteristics, if they are not precise the assigned type may also be incorrect.

Before explaining its operation, it is necessary to establish a series of definitions to be used in the rest of the document.

- **Individual Defect:** Each surface defect is called an individual defect. In Figure 4, each polygon represents a different individual defect. The relevant characteristics of each individual defect can be grouped into two categories: spatial and morphological. Spatial characteristics determine its position within the surface of the web material (both transversal and longitudinal coordinates). Morphological characteristics define its shape. These characteristics, provided by the inspection system, are its area (in pixels), its type and the length and width of its bounding box.

- **Lost Defect:** Inspection systems usually operate in real time, which can cause them to return inaccurate data. Thus, the characteristics of the detected individual defects can be inaccurate or an individual defect may even go undetected. These individual defects, which do exist on the surface of the inspected material but were not detected by the inspection system, are called lost defects.

- **Periodical Defect:** A set of individual defects on the same product that have been generated periodically by the same defective roll. In order to be considered a periodical defect, all individual defects within the set must fulfill the following conditions:

  ✓ All individual defects must be located in the same transversal position.

  ✓ All individual defects must have the same morphological characteristics, as they have been generated by the same flaw in the same defective roll.

  ✓ The separation between two consecutive individual defects must be constant. This separation is the period length of the periodical defect.

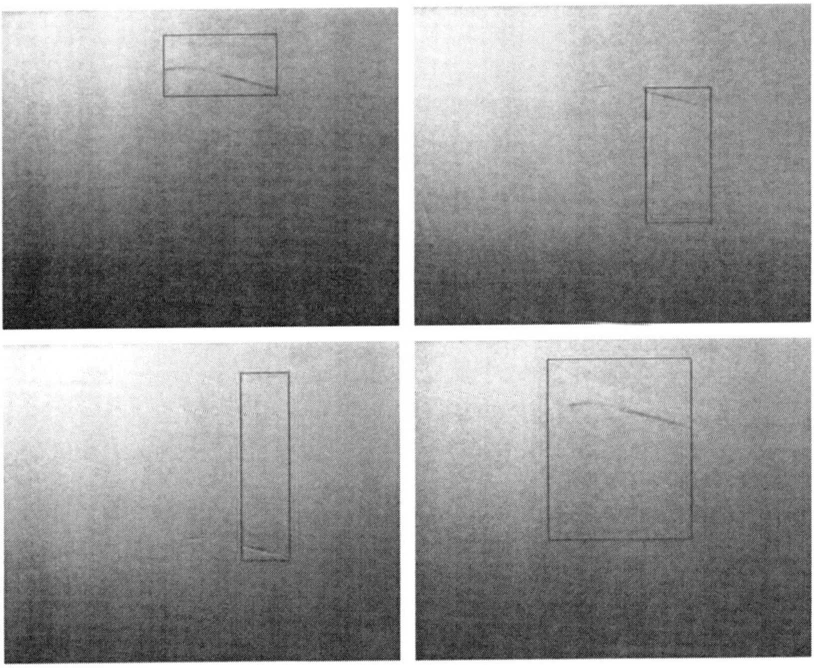

Figure 3. Inaccurate input data.

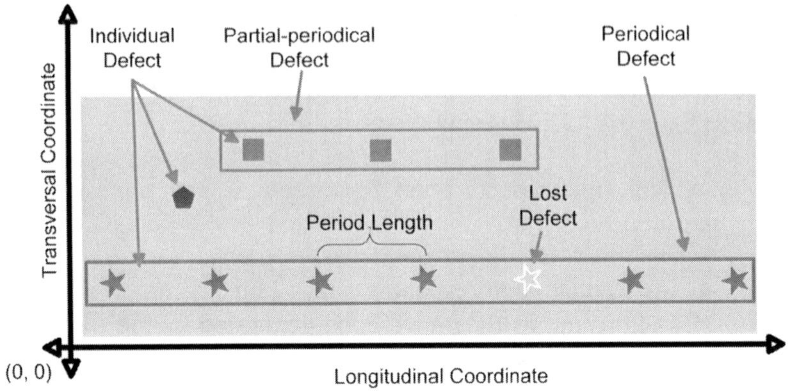

Figure 4. Defect definitions.

Although this is the definition of an ideal periodical defect, the uncertainty in the input data must be taken into account. This means that the transversal positions, dimensions and areas of the individual defects that form a periodical defect may be slightly different. Furthermore, the distance between two consecutive individual defects may not be exactly constant and lost defects may occur. Similarly, the type assigned to these individual defects may not be the same. All stars in Figure 4 form a set which meets the above conditions, so it can be considered a periodical defect.

- **Partial-periodical Defect:** A set that meets the conditions of periodical defect but is formed by a low number of individual defects is called a partial-periodical defect. Sometimes an object attached to a roll that is generating defects periodically can come off by itself. In these cases, the number of defects generated may be small. These cases are not of our interest, since the roll is not defective. For this reason, periodical defects with a low number of individual defects should not be considered as such and, therefore, should not be detected. This is the case of the individual defects represented by squares in Figure 4.

3.1. Solution Representation

Usually, inspection systems provide a large amount of information about each individual defect. In the proposed approach, the only information needed about each individual defect is its position (defined by its transversal and longitudinal coordinates), its area (in pixels), its dimensions (length and width) and its type (hole, scratch, roll mark, *etc.*). Individual defects will be represented by a tuple whose members correspond to each one of the characteristics mentioned above. Each one will be represented using one of these tuples.

$$id = \{tr, lo, ar, le, wi, ty\} \quad (1)$$

where *tr* and *lo* are the transversal and longitudinal coordinates of the individual defect, *ar* represents its area, and *le*, *wi* and*ty* its length, width and type respectively.

Since the aim of this study is the detection of periodical defects, an abstraction to represent them is also necessary. As periodical defects are composed by a set of individual defects generated by a defective roll periodically, they can be considered as a periodical pattern. This pattern can be defined by the following features.

$$pd = \{n, u, h, p, t, a, l, w, s\} \quad (2)$$

where *n* is the number of individual defects that belong to the periodical defect, *u* is the number of undetected individual defects, *h* is the highest number of consecutive undetected individual defects, *p* is its period length, *t* is its transversal position and *a*, *l* and *w* are the average area (in pixels), length and width of the individual defects that belong to the periodical defect. The feature *s* represents a set that stores the individual defects that make up the periodical defect (feature*n* is the cardinal of this set). Each periodical defect is characterized using one of these tuples.

In the remainder of the paper, the characteristic *X* of an individual defect is referred to by *id.X* and the feature *Y* of a periodical defect is referred to by *pd.Y*. The ith element of the set of a periodical defect is denoted by $pd.s.id_i$. All individual defects within the set are ordered by their longitudinal coordinate. Thus, the individual defect within a periodical defect whose longitudinal coordinate is lowest is denoted by $pd.s.id_1$, and the individual defect whose longitudinal coordinate is greatest is denoted by $pd.s.id_{pd.n}$. Thereby, every pair of elements within the set of a periodical defect must meet Equation (3).

$$pd.s.id_i.lo > pd.s.id_j.lo \quad \forall i, j \;\; i > j \tag{3}$$

As mentioned at the beginning of this section, the characteristics of individual defects have been taken with uncertainty. For this reason, all the individual defects that make up a periodical defect will not be located in the same transversal position, due to this uncertainty (although the transversal coordinates will not be equal, they will be very similar). The same situation occurs with the area, width and length. Thus, features *t*, *a*, *l* and *w* shall contain the average value of the characteristics *tr*, *ar*, *le* and *wi* of the individual defects that integrate the periodical defect, as defined in Equations (4–7)respectively.

$$pd.t = \frac{1}{pd.n} \sum_{i=1}^{pd.n} pd.s.id_i.tr \tag{4}$$

$$pd.a = \frac{1}{pd.n} \sum_{i=1}^{pd.n} pd.s.id_i.ar \tag{5}$$

$$pd.l = \frac{1}{pd.n} \sum_{i=1}^{pd.n} pd.s.id_i.le \tag{6}$$

$$pd.w = \frac{1}{pd.n} \sum_{i=1}^{pd.n} pd.s.id_i.wi \tag{7}$$

The value of the feature *p* contains the average distance between two consecutive individual defects belonging to the periodical defect (taking into account the undetected individual defects).

$$pd.p = \frac{pd.s.id_{pd.n} - pd.s.id_1}{pd.n + pd.u - 1} \tag{8}$$

Finally, the value of the feature *h* (the highest number of consecutive undetected individual defects) can be calculated as shown in Equation (9).

$$pd.h = \frac{\max\limits_{1 \leq i \leq pd.n-1} (pd.s.id_{i+1}.lo - pd.s.id_i.lo)}{pd.p} - 1 \quad (9)$$

This is the way in which each periodical defect is represented. The output provided by the clustering method used by the proposed sensor system consists of a collection of these tuples.

3.2. Detection of Periodical Defects

In this section, the way that periodical defects are detected is discussed. Initially, Section 3.2.1 shows how the initial values of a periodical defect are established. The initial solution is completed by adding individual defects that meet the conditions of a periodical defect using a clustering algorithm described in Section 3.2.2.

3.2.1. Initial Solution

The amount of time available for detection may be limited. Thus, the search should examine the transversal coordinates most likely to have periodical defects first (the position of an individual defect is defined by its longitudinal and transversal coordinates, but the position of a periodical defect is defined by its transversal position only). To do this, a histogram representing the number of individual defects located at each transversal position is calculated, as shown in Figure 5. The transversal position containing the maximum of the histogram is chosen to start the search. In order to start a search at that transversal position, a periodical defect is created. Its tuple is initialized including an individual defect located at the transversal position in which periodical defects are being sought (any of them). In addition, before starting the search, the theoretical period length (*tpl*) of the periodical defects to be detected should be established (the theoretical period length of a roll could be defined as the period of the periodical defects generated by it). Generally, the theoretical period length of a roll is its perimeter, but this depends on the production process of each material. If multiple rolls are used in the production process, a search should be performed for each one of them (using their theoretical period lengths). Thus, the initial values of the features of the periodical defect are those shown in Equation (10).

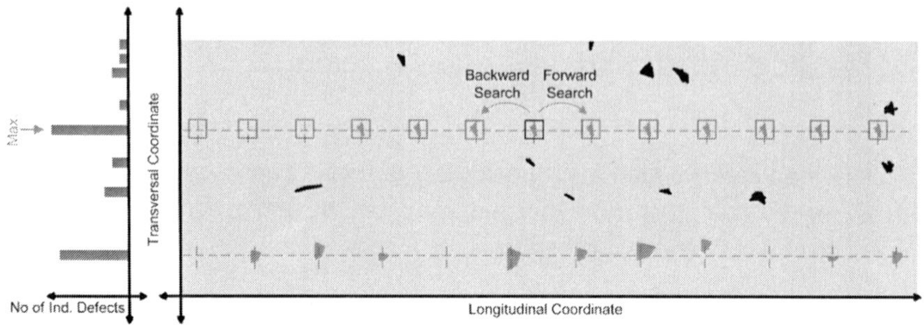

Figure 5. Clustering algorithm.

$$pd = \{1, 0, 0, tpl, s.id_1.tr, s.id_1.ar, s.id_1.le, s.id_1.wi, s\} \quad (10)$$

3.2.2. Clustering Algorithm

Once the transversal position, in which a periodical defect is going to be sought, has been selected and an initial solution (a periodical defect with only one individual defect included in its set and located at that transversal position) has been created, applying a clustering algorithm, other individual defects are added to the set. The clustering algorithm aims to find individual defects that match the one included in the initial solution and include them in the set, with the objective of obtaining the features of the periodical defect located in the transversal position determined by the histogram (in the case where a periodical defect actually exists there). These individual defects are those that meet the conditions of being a periodical defect, explained above. To perform this task, the theoretical period of each roll involved in the production process should be taken into account.

Figure 6 shows the main steps of the clustering method. Accordingly, once the initial solution has been created and the clustering algorithm has been executed, periodical defects placed at other transversal positions are sought. To do this, the histogram is calculated again, without taking into account the individual defects included in a periodical defect previously detected. An initial solution is created for that transversal position and the clustering method is applied again. These steps are repeated until all transversal positions have been checked or the time available for performing the detection is exhausted.

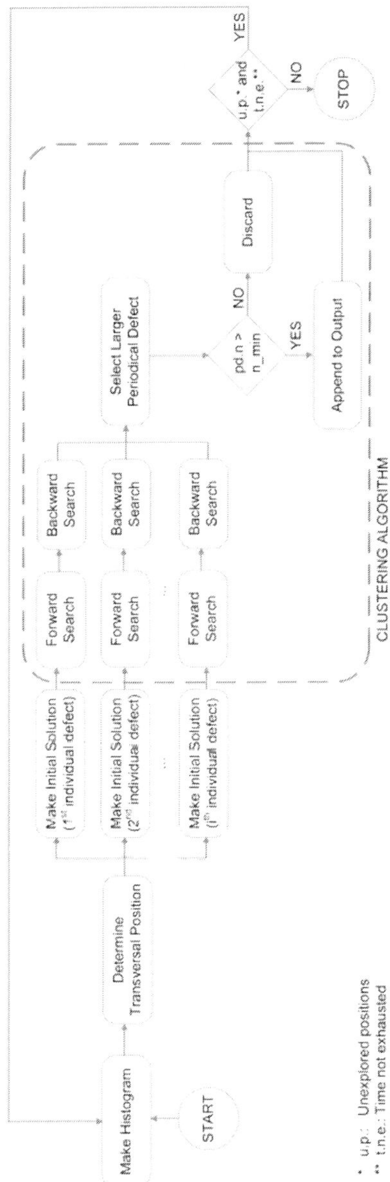

Figure 6. Steps of the clustering method.

In Figure 5 an example with two periodical defects is shown. One of them, shown at the top of the figure, is located at the transversal position in which the histogram has its maximum. The initial solution must contain an individual defect located in that transversal position (in the example, the individual defect that is within a black square). The clustering algorithm should look for individual defects that match it. These must be located in the same transversal position; they must be characterized as the same type and have the same area, length and width. Moreover, their longitudinal position must be consistent with the longitudinal position of the individual defect included in the initial solution, considering the theoretical period length of the rolls involved in the production process (the initial value of the feature *p* of the periodical defect). Assuming the case where the initial periodical defect is the one shown in Equation (10) and the individual defect included in the initial solution (designated as *iid* and at this point, $pd.s.id_1$ is the one shown in Equation (11), individual defects whose position fits with it are those located in ($iid.tr, iid.lo \pm N \cdot pd.p$), where *N* is a natural number. These positions are represented in Figure 5 by dashed blue lines.

$$iid = \{tr, lo, ar, le, wi, ty\} \tag{11}$$

This approach works properly only in ideal situations, where all individual defects are located at the same transversal position, the longitudinal separation between them is constant, there are no lost defects, and their dimensions (length, width and area) are identical. As already remarked, input data may be inaccurate, so a real periodical defect is most similar to that shown at the bottom of Figure 5, where the individual defects may have different length, width, area, transversal position and type. In addition, their separation may not be constant (in longitudinal direction) and some individual defects may have not been detected. Thus, various tolerances should be used when determining where other individual defects of the periodical defects should be located and when deciding whether an individual defect matches the features of the periodical defect or not. Related to the position, it is necessary to establish two tolerances: one for the transversal position and another for the longitudinal position. The transversal tolerance (hereinafter t_tol) establishes the maximum allowable difference between the transversal position of an individual defect and the feature *t* of the periodical defect where it aims to be included (in *mm*). The longitudinal tolerance (hereinafter p_ratio) establishes the maximum allowable difference between the longitudinal position of an individual defect candidate to be included in a periodical defect, and the longitudinal position where it should be located. This tolerance is defined as a percentage of the *p* feature of the periodical defect. Thereby, individual defects whose position should be accepted as similar to the individual defect defined in Equation (11), taking into account these tolerances, are those located in (*iid.tr* ± *t_tol, iid.lo* ± *N* ·*pd.p* ± *pd.p* · *p_ratio*). These areas are represented inFigure 5 by red squares and they constitute the search space of the algorithm. The width and length of these squares are defined by t_tol and p_ratio respectively.

The search space is explored in two steps: the forward search and the backward search. Both start from the position of the initial defect, located in (*iid.tr, iid.lo*). When performing the forward search, the algorithm starts checking whether an individual defect is located in (*iid.tr* ± *t_tol, iid.lo* ± *pd.p* ± *pd.p* · *p_ratio*). If there is an individual defect within that area (candidate individual defect, or *cid*), its characteristics should be checked. That individual defect is only included in the periodical defect if it meets the following conditions:

$$cid.ar \in [\frac{pd.a}{a\_ratio/100}, pd.a \cdot a\_ratio/100] \quad (12)$$

$$cid.le \in [\frac{pd.l}{l\_ratio/100}, pd.l \cdot l\_ratio/100] \quad (13)$$

$$cid.wi \in [\frac{pd.w}{w\_ratio/100}, pd.w \cdot w\_ratio/100] \quad (14)$$

where *a_ratio, l_ratio* and *w_ratio* are the morphological tolerances. They establish the maximum allowable difference between the area, length and width of an individual defect candidate to be included in a periodical defect, and the area, length and width that it should have: *pd.a, pd.l* and *pd.w* respectively. These tolerances are defined as a percentage of the *a*, *l*and *w* feature of the periodical defect. In addition, before starting the execution of the clustering algorithm, individual defect types that are allowed to form the periodical defects must be established. The type of the candidate individual defect must be one of the allowed types.

If the individual defect meets these conditions, it is included in the periodical defect (if there is more than one individual defect within the area that meets the above conditions, the chosen defect is the one closest to the ideal position).

Each time a candidate individual defect is included in the periodical defect, its *n* feature is incremented by one and the features *t*, *a*, *l*, *w* and *p* are updated using Equations (4–8) respectively.

This process is repeated to check the other squares (search areas) in the search space, *i.e.*, areas centered at the points (*iid.tr, iid.lo* + *pd.p*), (*iid.tr, iid.lo* + 2 · *pd.p*), (*iid.tr, iid.lo* + 3 · *pd.p*) and so on, until the end of the web material is reached.

Note that the longitudinal positions of the points above are calculated using *pd.p* instead of the theoretical period length since it may have been estimated with low precision. However, *pd.p* contains the actual average distance between individual defects. Thus, the error in the case that the theoretical period length has not been calculated with sufficient accuracy is minimized. In order to further minimize the error in these calculations, both the transversal and the longitudinal position of these points should not be determined as a function of *iid.tr* and *iid.lo* but by the average transversal position (*pd.t*) and the longitudinal position of the last defect inserted in the set ($pd.s.id_{pd.n}$). In this way the area to be checked until the end of the web material is reached is defined by Equation (15).

$$(pd.tr \pm t\_tol, pd.s.id_{pd.n}.lo + pd.p \pm pd.p \cdot p\_ratio) \quad (15)$$

Each time an individual defect is introduced into the set, it is the one whose longitudinal position is higher, *i.e.*, the latest individual defect inserted into the set is always $pd.s.id_{pd.n}$ during the forward search. In this way the area defined byEquation (15) changes every time an individual defect is introduced into the set, defining the area where the next one must be sought.

If there are no defects within the area defined by Equation (15) (or none satisfies the conditions Equations (12,13,14), the feature *pd.u* is increased by one and the feature *pd.h* is updated by Equation (9). In addition, the search area must be defined taking into account this situation. To do this, an auxiliary variable α is defined, with a default value of 1. Each time an individual defect is inserted into the set, the value of α is reset to 1. Like *pd.u*, when there are no individual defects within the search area its value is incremented by one. Thus, the search can continue even when individual defects were not inserted into the set, as shown in Figure 7. The general expression for defining the search area during the forward search is shown in Equation (16).

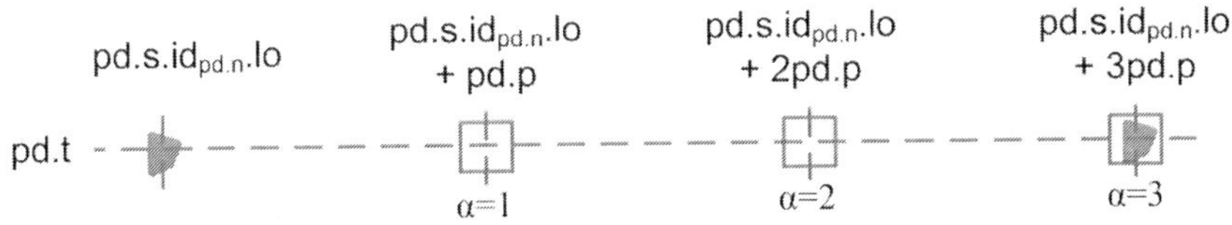

Figure 7. Forward search with lost defects.

$$(pd.tr \pm t\_tol, pd.s.id_{pd.n}.lo + \alpha pd.p \pm pd.p \cdot p\_ratio) \quad (16)$$

When many individual defects exist in the web material inspected, some of them may be located in close transversal positions and have similar morphological characteristics by chance. These defects could be incorrectly clustered as a periodical defect. To avoid this situation, it is necessary to set two thresholds, defined as follows.

- **Size Threshold (n_min):** Sets the minimum number of individual defects that a periodical defect must have to avoid being considered a partial-periodical defect.
- **Undetected Defects Threshold (max_skips):** Maximum allowable number of individual defects undetected consecutively.

By using size thresholds, the probability that a set of defects, such as those mentioned above, are classified as a periodical defect is minimized. Also, it establishes the difference between partial-periodical and periodical defect. In addition, the undetected defects threshold can be used by technicians to set the highest number of lost defects they deem acceptable (that is, the maximum allowed value for the parameter *h* of a periodical defect and the highest value that can take the variable *α*). If the technician considers it unnecessary to stop the production line when the individual defects that form the periodical defects are subtle, a small value for *h* can be established (subtle defects are more difficult to detect, and therefore many of them may go undetected). On the contrary, if it is considered necessary to stop the line whenever periodical defects are being generated (even though they are very subtle) a large value for *h* can be established. During both searches, forward and backward, Equation (17) must be fulfilled:

$$\alpha < max\_skips + 1 \quad (17)$$

During the forward search, when the end of the web material is reached or Equation (17) ceases to be met, it ends and the backward search begins. The backward search looks for individual defects to be included into the set from the position where the first individual defect is located ($pd.s.id_1.tr$, $pd.s.id_1.lo$) to the beginning of the web material. The manner in which individual defects are sought is analogous to the forward search. The area within which the next individual defect is sought is defined by an expression similar to that used during the forward search, but in this case it must be taken into account that the last defect included in the set is the one that is located in a lower longitudinal position, *i.e.*, $pd.s.id_1$. Whenever an individual defect is included in the set, it will become the new $pd.s.id_1$ (since its longitudinal position is the lowest). Thus, the search area for the backward search is defined by Equation (18).

$$(pd.tr \pm t\_tol, pd.s.id_1.lo - \alpha pd.p \pm pd.p \cdot p\_ratio) \quad (18)$$

When the beginning of the web material is reached or Equation (17) ceases to be met, the backward search stops and the whole search ends.

A very important aspect that must be taken into account is that the individual defect selected to form the initial solution greatly affects the search, since the values of their characteristics *iid.tr, iid.ar, iid.wi* y *iid.le* are assigned to features *pd.t, pd.a,pd.w* y *pd.l* of its periodical defect respectively at the beginning of the search. Thus, the clustering algorithm includes in the periodical defect only those individual defects whose characteristics are similar to those of the individual defect included in the initial solution. Therefore, if this individual defect was detected by the vision system incorrectly (e.g., assigning a value to its dimensions and area much larger than they should be) the clustering algorithm will not work properly, since it will include only individual defects similar to it (whose dimensions and area are also incorrect), as shown in Figure 8. In this figure, the individual defect chosen to form the initial solution is surrounded by a black square, and individual defects included in the periodical defect by the clustering algorithm are surrounded by red squares. The clustering is better in the cases where the individual defect included in the initial solution is more like the majority of individual defects that form the periodical defect. Since the individual defect most appropriate to form the initial solution can not be determined before running the algorithm, it must receive as many initial solutions as individual defects are located at the transversal position determined by the histogram (see Figure 6). For each of them, both forward and backward searches are applied. The largest periodical defect (*i.e.*, the one whose feature *n* is greater) is the one chosen to be part of the output, but only if its feature *n* is greater than the size threshold.

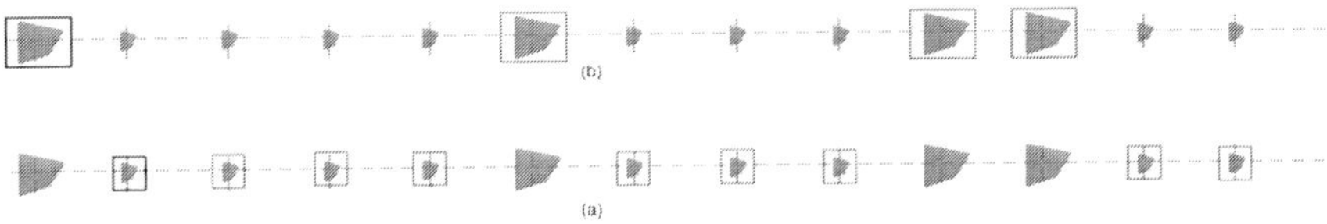

Figure 8. Differences in detection depending on the initial solution.

3.3. Parameter Settings

The tolerances and thresholds constitute the configuration parameters of the clustering algorithm, adjusting its behavior. Thus, their values must be set appropriately taking into account the type of web material inspected and the inspection system used. If the uncertainty introduced by the inspection system is small, the parameters should be restrictive, and if it is large, the parameters should be more permissive.

For each case in which the clustering method is going to be used, the optimal value of each parameter must be set. To do this, the output provided by the sensor must be assessed. Thus, the parameter configuration whose output has the best assessment can be considered the best. A method to perform this assessment using discrepancy based metrics is proposed in Bulnes *et al.* [11]. In order to determine the optimal parameter values in this way, it is necessary to have several examples of the web material for which the proposed sensor is going to be used, where the inspection task is carried out by the same system that provides the input data. For each one of these examples, an expert technician must make the detection of periodical defects manually a priori. This detection will be considered perfect and used as reference. The assessment method accounts for the differences between the output provided by the proposed sensor and the reference, and returns a metric value within the closed interval [0, 1] (the higher the value, the fewer differences). Specifically, the F-Measure metric was chosen for this purpose [29]. The selected parameter configuration is the one that provides a higher average metric value.

4. PRACTICAL APPLICATION: DETECTION OF PERIODICAL DEFECTS IN HOT STEEL STRIPS

A practical application has been taken into consideration to assess the functioning of the sensor presented in this paper: hot steel strips. It has been chosen because it is a case in which periodical defects are difficult to detect. This is because there are several rolls involved in the production process, the number of surface defects is usually high and the input data are very noisy.

4.1. Industrial Context

In this application, the generation of periodical defects occurs in the finishing mill (Figure 9), where hot steel is stretched and flattened by several rolls, exerting great pressure on steel until it reaches the form of a strip of great length. During the rolling process, the strip has to pass through seven stands. Each one consists of four rolls: two backup rolls and two work rolls. The backup rolls transmit force to the work rolls, which are those in direct contact with the steel. For this reason, the work rolls are those that can generate periodical defects on the surface of the strips. The surface of the work rolls must be completely smooth to obtain a perfect rolling. Besides the typical problems on rolls (cracks or attached objects), in this particular case they become worn very quickly [30], which makes the inspection even more important so as to replace the malfunctioning rolls as early as possible.

As discussed in Section 3.2.1, the theoretical period length of the rolls involved in the production process must be known before starting the detection. Thus, the time required to perform the detection can be reduced by searching only periodical defects with such periodicities, instead of iterating through a range of possible values as made by Traxler *et al.* [25]. In our practical application, the theoretical period length does not correspond to the perimeter of the rolls since the length of the material varies, but it can be calculated easily. The forces applied during rolling produce friction on the strip, which is lower in the transversal direction than in the forward direction of the strip, cause the stretching of the strip in a forward direction and not sideways. Figure 10 shows an example in which a defective roll generates a periodical defect whose period length is initially L (the perimeter of that roll) when the thickness of the strip is S. Passing through the next stand, the thickness of the strip is reduced to $S/2$. Since the volume of the steel remains constant, the length of the steel increases proportionally, causing the separation between two consecutive individual defects within the periodical defect to increase to $2L$. Considering the separation between each pair of work rolls of the same stand and their radii, the theoretical period of each cylinder can be calculated by Equation (19) [31].

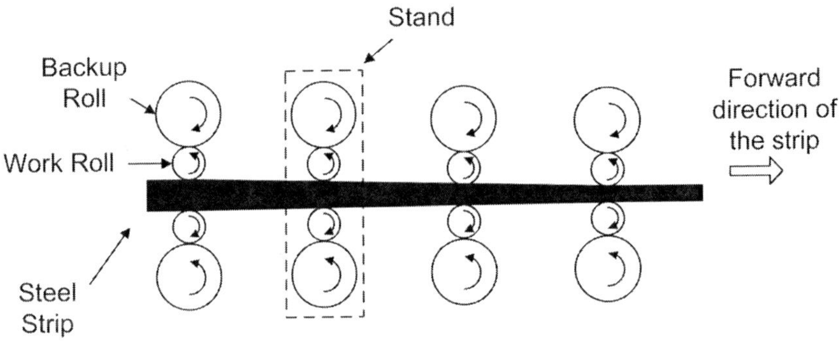

Figure 9. Steel strip at the finishing mill.

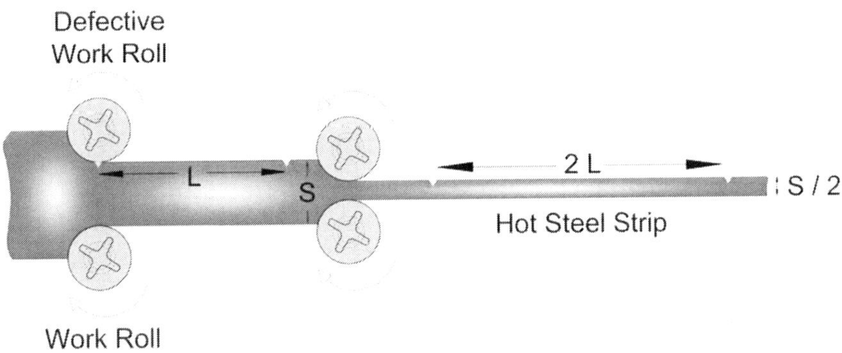

Figure 10. Relation between the reduction of thickness and the elongation of the strip.

$$p_i = (2\pi) r_i \prod_{j=i+1}^{n} \frac{1}{1-b_j} \tag{19}$$

where n is the number of stands in the finishing mill, p_i is the theoretical period length of the work rolls of the ith stand andr_i is its radius. The reduction factor b_j measures the thickness difference of the strip due to passing through the jth stand.

$$b_j = \frac{h_{j-1} - h_j}{h_{j-1}} \tag{20}$$

where h_k is the thickness of the steel strip after passing through the kth stand.

4.2. Experimentation

To perform the experimental phase, 36 strips produced in the ArcelorMittal Avilés factory (Spain) have been used. In order to achieve representative results, strips with different characteristics have been chosen. The number of periodical defects in each strip varies from 1 to 5, where the number of individual defects that form them varies from 20 to 250 and the number of individual defects in each strip varies from 80 to nearly 5000. They are also strips of different types of steel, different thicknesses and dimensions.

All these strips have been inspected during their production process using an inspection system by Parsytec [32] which is able to both detect and classify individual defects and also perform the detection of periodical defects. This system uses CCD cameras that take pictures in grayscale with a resolution of 768 × 494. Those images are overlapped to ensure complete coverage. Depending on the side and the camera, the overlap varies from 5% to 15%. The separation between the cameras and the strip is 2 meters on the top side and 3.67 meters on the bottom side. Since the steel strips are moving very fast (between 11 and 17 meters per second) exposure time of the cameras is low: 100μs. The strips are illuminated by xenon strobes emitting 120 flashes per second (12μs of duration each). The detection of periodical defects by the detection method proposed has been entirely written in C++ using *Microsoft Visual Studio 2008*. These experiments were performed on a computer that uses an *Intel Core2 Quad Processor Q9400*, taking 7 seconds (on average) to analyze a full strip.

The data provided by the inspection system about the detection and classification of individual defects is used as input data, while the periodical defect detection is used for comparison with the output provided by the proposed sensor, in order to check whether it represents an improvement with respect to an important system used worldwide.

The first step in applying the proposed method is to determine the optimal parameter configuration of the clustering method for this case using that inspection system. Then, this configuration is used to perform the testing phase. In order to avoid obtaining an overfitted configuration, the strips selected for experimentation have been divided into two different groups of equal size (18 strips each). The first one is used to obtain the optimal parameter configuration and the second one to perform the testing.

Using a full-factorial experimental design executed by a grid composed of 60 computers, the optimal parameter configuration was obtained. The one that provides better average results for the first set of strips is shown in Table 1.

Table 1. Optimal configuration.

Parameter	Value
n_min	18
max_skips	20
a_ratio	100
w_ratio	100
l_ratio	100
t_tol	56
p_ratio	7

The values obtained for the tolerances applied to the morphological characteristics of individual defects are very high in the optimal configuration due to the low accuracy under which they were obtained in this particular case. This means that two different individual defects are considered morphologically similar if the area, length and width of one of them is 100 times greater than the other. Practically this means that the morphological characteristics of the individual defects are not taken into account in this case, where the detection of periodical defects are performed based on the position and type of the individual defects only. Moreover, the value of the parameter *max_skips* is also very large because the occurrence of lost defects is very common.

The last step of the experimentation is to make the detection of periodical defects for the second set of strips using the proposed sensor, applying the obtained parameter configuration for the clustering method. The results are compared with those provided by the Parsytec system in order to assess whether the behavior of the proposed sensor is good enough to work in a real industrial environment.

4.3. Results

The obtained results after using the proposed sensor on the set of test strips is shown in Figure 11 with the results provided by the Parsytec system for that same set. The numerical data are also listed in Table 2. These results show a clear improvement, since the output provided by the proposed sensor has obtained a better rating than the Parsytec system for all strips. Table 3 shows the average metric values provided by each system.

These results suggest that the quality of the output of the proposed sensor is better than that provided by the Parsytec system. To verify this assertion, a statistical test needs to be applied. The first step is to check whether the data are normally distributed. To do this we used the Shapiro–Wilk test (All statistical calculations were performed using R), concluding that the hypothesis of normality for the data provided by the proposed method can be rejected (p-value = 0.0011), but not for data provided by the Parsytec system (p-value = 0.1222). Due to the lack of normality, a one-sided Wilcoxon test for paired data should be used. The distribution D is defined as the difference between the metric value provided by the proposed method and system Parsytec for each strip. Thus, the hypotheses to be tested using this method are defined as:

$$\begin{cases} H_0 : \text{the median of the distribution D is less than or equal to 0} \\ H_1 : \text{the median of the distribution D is greater than 0} \end{cases}$$

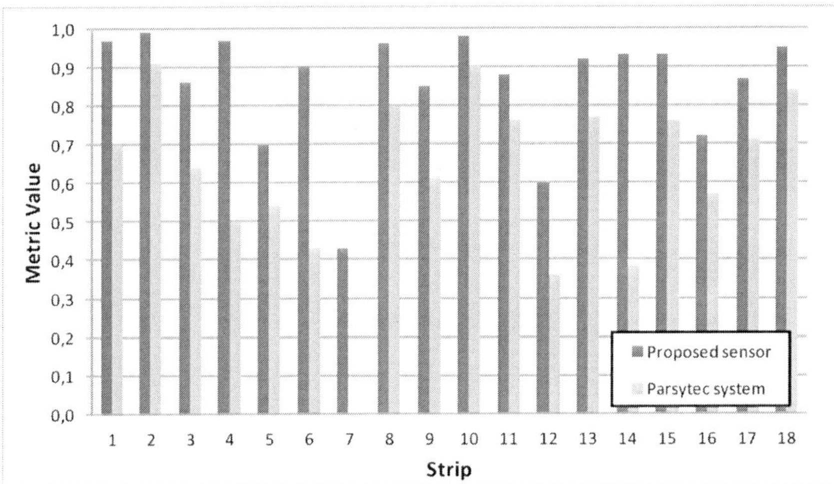

Figure 11. Comparison between the proposed sensor and the Parsytec system.

Applying the Wilcoxon test, a *p*-value of 0.00011 is obtained, allowing the rejection of the null hypothesis with confidence. Thus, it can be concluded that the results obtained by the proposed sensor are statistically superior to those provided by the Parsytec system.

5. CONCLUSIONS

This paper has presented a sensor system based on computer vision capable of detecting periodical defects in web materials during their production. This kind of defects requires an early detection, as the defects can greatly degrade the quality of the produced material. Therefore, the clustering method used by the sensor is fast and enough to be able to provide the results within a few seconds without using a powerful computer.

Table 2. Metric values obtained for each strip.

Strip	1	2	3	4	5	6	7	8	9	10	11	12	13	14	15	16	17	18
Proposed sensor	0.97	0.99	0.86	0.97	0.70	0.90	0.43	0.96	0.85	0.98	0.88	0.60	0.92	0.93	0.93	0.72	0.87	0.95
Parsytec system	0.70	0.91	0.64	0.50	0.54	0.43	0.00	0.80	0.61	0.90	0.76	0.36	0.77	0.38	0.76	0.57	0.71	0.84

Table 3. Average F-Measure metric values.

	Average Metric Value
Proposed method	0.86
Parsytec system	0.62

All steps of the clustering method have been detailed. It is based on the position and the morphology of the surface defects to determine whether a set of them may have been generated periodically by a defective roll. In order to function properly even with noisy input data, a tolerance is applied to each of the defect characteristics that are taken into account. Thus, by changing the value of each tolerance, the behavior of the proposed method can be adapted to operate optimally for the web material to be inspected.

To verify the correct operation of the sensor system, it was applied to the detection of periodical defects in hot steel strips. The optimal configuration for this particular case was determined and then used in real steel strips. The results were compared with those provided by an inspection system that is widely used for the detection of periodical defects, which is considered the state-of-the-art in this field. The comparison proves the success of the proposed sensor and its capacity to detect periodical defects in any industry.

ACKNOWLEDGMENTS

This research has been partially supported by the project TIN2011 24903 of the Spanish National Plan of Research and Development and ArcelorMittal Spain. It has been conducted with the assistance of Statistical Consulting Unit of the University of Oviedo. The authors would like to thank the technicians and engineers of ArcelorMittal Asturias for their help during tests and for their gracious time. They also would like to thank the anonymous reviewers for their helpful comments and suggestions.

REFERENCES

1. Malamas, E.N.; Petrakis, E.G.M.; Zervakis, M.; Petit, L.; Legat, J.D. A survey on industrial vision systems, applications and tools. *Image Vis. Comput.* **2003**, *21*, 171–188.

2. Zhang, Z.F.; Gao, Z.; Liu, Y.Y.; Jiang, F.C.; Yang, Y.L.; Ren, Y.F.; Yang, H.J.; Yang, K.; Zhang, X.D. Computer vision based method and system for online measurement of geometric parameters of train wheel sets. *Sensors* **2012**, *12*, 334–346.

3. Adamo, F.; Attivissimo, F.; Nisio, A.D. Calibration of an inspection system for online quality control of satin glass.*IEEE Trans. Instrum. Meas.* **2010**, *59*, 1035–1046.

4. Huang, W.; Kovacevic, R. A laser-based vision system for weld quality inspection. *Sensors* **2011**, *11*, 506–521.

5. Ge, W.; Zhao, H.; Li, X. Gyroscope pivot bearing dimension and surface defect detection. *Sensors* **2011**, *11*, 3227–3248.

6. Navarro, P.; Iborra, A.; Fernandez, C.; Sanchez, P.; Suardiaz, J. A sensor system for detection of hull surface defects.*Sensors* **2010**, *10*, 7067–7081.

7. Telljohann, A. Chapter 2, Introduction to Building a Machine Vision Inspection. In *Handbook of Machine Vision*; Wiley-VCH Verlag GmbH & Co. KGaA: Weinheim, Germany, 2007; pp. 35–71.

8. Brzakovic, D.; Vujovic, N. Designing a defect classification system: A case study. *Pattern Recog.* **1996**, *29*, 1401–1419.

9. Chang, C.A.; Su, C.T. A comparison of statistical regression and neural network methods in modeling measurement errors for computer vision inspection systems. *Comput. Ind. Eng.* **1995**, *28*, 593–603.

10. Bulnes, F.G.; Garcia, D.F.; Usamentiaga, R.; Molleda, J. A Technique for Clustering Individual Defects from Images of Steel Strips in Periodical Defects. Proceedings of the 12th IAPR Conference on Machine Vision Applications, Nara, Japan, June 2011; pp. 283–286.

11. Bulnes, F.G.; Usamentiaga, R.; Garcia, D.F.; Molleda, J.; Rendueles, J.L. Vision-based technique for periodical defect detection in hot steel strips. *IEEE Ind. Appl. Mag.* **2012**. in press.

12. Bulnes, F.G.; Usamentiaga, R.; Garcia, D.F.; Molleda, J. Detection of Periodical Patterns in the Defects Identified by Computer Vision Systems. Proceedings of the IEEE 11th International Conference on Intelligent Systems Design and Applications, Cordoba, Spain, November 2011; Volume 1. pp. 301–306.

13. Bissessur, Y.; Martin, E.; Morris, A. Monitoring the performance of the paper making process. *Control Eng. Pr.* **1999**, *7*, 1357–1368.

14. I'Anson, S. Identification of periodic marks in paper and board by image analysis using two-dimensional fast Fourier transforms. *Tappi J.* **1995**, *78*, 113–119.

15. Bernie, J.P.; Trepanier, R. A New Technique to Characterize Periodic and Pseudo-Periodic Marks Within a Sheet of Paper. Proceedings of the 2007 TAPPI Papermakers and PIMA International Leadership Conference, Jacksonville, FL, USA, March 2007.

16. Bernie, J.; Trepanier, R. A new technique for measuring periodic patterns within a paper sheet. *Pulp Pap. Can.* **2009**,*110*, 39–42.

17. Chaplin, D. Image analysis: A powerful tool to pinpoint defects and gauge print quality. *Pap. Age* **1998**, *114*, 30–31.

18. Kumar, A. Computer-vision-based fabric defect detection: A survey. *IEEE Trans. Ind. Electron.* **2008**, *55*, 348–363.

19. Toba, E. Determination of autocorrelation function of woven fabrics by using laser speckle. *Text. Res. J.* **1980**, *50*, 238–244.

20. Drean, J.Y.; Harzallah, O.; Renner, M. Textile surfaces analysis and modeling based on statistical mmethods: Variance analysis and autocorrelation functions. *Text. Res. J.* **2010**, *80*, 1833–1845.

21. Kianiha, A.; Sheikhzadeh, M.; Semnani, D.; Dabirian, F. An investigation into the weft insertion periodic errors in fabric using an image analysis method. *J. Text. Inst.* **2010**, *101*, 457–462.

22. Duan, X.; Duan, F.; Han, F. Study on Surface Defect Vision Detection System for Steel Plate Based on Virtual Instrument Technology. Proceedings of the 2011 International Conference on Control, Automation and Systems Engineering CASE 2011, Singapore, Singapore, July 2011.

23. Lee, K.S.; Kim, K. Surface quality improvement by SDD in POSCO. *SEAISI Q.* **2003**, *32*, 58–58.

24. Shigeno, M.; Yamauchi, K.; Amanuma, Y. *Development of Surface Detector for No. 3 Tin Temper Mill (3TMP)*; JFE Technical Report; JFE Holdings Inc.: Tokyo, Japan, 2007; pp. 64–69.

25. Traxler, G.; Klarner, J.; Huelble-Koenigsberger, G. Broken roll detection, application, algorithm and its basic principles of sensing. *Proc. SPIE* **2005**, *5679*, 146–155.

26. Park, C.; Choi, S.; Won, S. Vision-based inspection for periodic defects in steel wire rod production. *Opt. Eng.* **2010**, *49*.

27. Oerley, H.; Bauereiss, U.; Strobl, F. In-Line Inspection of Base Film and Its Coating. Proceedings of the TAPPI European PLACE Conference 2, Budapest, Hungary, May 2009; Volume 2. pp. 1610–1649.

28. Caron, J.; Duvieubourg, L.; Orteu, J.; Revolte, P. Real-time inspection of preweathered Zinc based on computer vision.*Proc. SPIE* **1996**, *2665*, 270–281.

29. Van Rijsbergen, C.J. *Information Retrieval*, 2nd ed.; Butterworths: Oxford, United Kingdom, 1979.

30. Cowling, P. A flexible decision support system for steel hot rolling mill scheduling. *Comput. Ind. Eng.* **2003**, *45*, 307–321.

31. Zhao, S.; Huang, H.; Wang, F.; Li, Y.; Wei, G. Estimating periodic length and defect coordinates from a series of integer values. *Meas. J. Int. Meas. Confed.* **2009**, *42*, 516–523.

32. Jeong, J.H.; Kalowitsch, M. State-of-the-art surface inspection technology with parsytec 5i. *SEAISI Q.* **2007**, *36*, 10–15.

Chapter 9

Multi-Threshold Algorithm Based on Havrda and Charvat Entropy for Edge Detection in Satellite Grayscale Images

Mohamed A. El-Sayed[1,2], Hamida A. M. Sennari[3]

[1]Department of Mathematics, Faculty of Science, Fayoum University, Al Fayoum, Egypt; [2]Assistant Professor of Computer Science, Taif University, Taif, KSA; [3]Department of Mathematics, Faculty of Science, Aswan University, Aswan, Egypt.

ABSTRACT

Automatic edge detection of an image is considered a type of crucial information that can be extracted by applying detectors with different techniques. It is a main tool in pattern recognition, image segmentation, and scene analysis. This paper introduces an edge-detection algorithm, which generates multi-threshold values. It is based on non-Shannon measures such as Havrda & Charvat's entropy, which is commonly used in gray level image analysis in many types of images such as satellite grayscale images. The proposed edge detection performance is compared to the previous classic methods, such as Roberts, Prewitt, and Sobel methods. Numerical results underline the robustness of the presented approach and different applications are shown.

KEYWORDS

Multi-Threshold; Edge Detection; Measure Entropy; Havrda & Charvat's Entropy

1. INTRODUCTION

Edge detection is a very important tool used in many applications of image processing to obtain information from the frames as a preparatory step to feature extraction and object segmentation. This phase detects outlines of an object and boundaries between objects and the background in the image [1]. The detection results benefit applications such as optical character recognition [2], infrared gait recognition [3,4], automatic target recognition [5], detection of video changes [6], and medical image applications [7].

Edge detection concerns localization of abrupt changes in the gray level of an image [8]. Edge detection can be defined as the boundary between two regions separated by two relatively distinct gray level properties [9]. The causes of the region dissimilarity may be due to some factors such as the geometry of the scene, the radio metric characteristics of the surface, the illumination and so on [10]. An effective edge detector reduces a large amount of data but still keeps most of the important feature of the image. Edge detection refers to the process of locating sharp discontinuities in an image [11,12].

Many operators have been introduced in the literature, for example, Roberts, Sobel and Prewitt [13-17]. Edges are mostly detected using either the first derivatives, called gradient, or the second derivatives, called Laplacien. Laplacien is more sensitive to noise since it uses more information because of the nature of the second derivatives.

Most of the classical methods for edge detection based on the derivative of the pixels of the original image are Gradient operators, Laplacien and Laplacien of Gaussian (LOG) operators [10]. Gradient based edge detection methods, such as Roberts, Sobel and Prewitts, have used two linear filters to process vertical edges and horizontal edges separately to approximate first-order derivative of pixel values of the image. Marr and Hildreth achieved this by using the Laplacien of a Gaussian (LoG) function as a filter [18]. To solve these problems, the study proposed a novel approach based on information theory, which is entropy-based thresholding. The proposed method is to decrease the computation time compared with Canny and LoG method. The results were very good compared with the well-known Roberts, Prewitt, and Sobel gradient results.

The outline of the paper is as follows. In Section 2, we have presented the classical edge detection methods that related to the paper. Image thresholding based on Havrda & Charvat's entropy is presented in Section 3. Section 4 describes the edge detection that was based on entropy. Section 5 illustrates the multi-threshold algorithm based on Havrda and Charvat entropy for edge detection. In Section 6, we have presented the effectiveness of proposed algorithm in the case of satellite grayscale images, and also we compared the results of the algorithm with several leading edge detection methods such as Roberts, Prewitt, and Sobel methods in the same section. Conclusions are presented in Section 7.

2. CLASSICAL EDGE DETECTION METHODS

Five most frequently used edge detection methods are used for comparison. These are: Gradient operators (Roberts, Prewitt, Sobel), Laplacian of Gaussian (LoG or Marr-Hildreth) and Gradient of Gaussian (Canny) edge detections [19,20]. People which would like to read about this subject are referred to [21-23] evaluation studies of edge detection algorithms according to different criteria. The details of methods as follows.

2.1. Roberts Edge Detector

The Roberts Cross operator performs a simple, quick to compute, 2-D spatial gradient measurement on an image as shown in Figure 1. It thus highlights regions of high spatial frequency which often correspond to edges. In its most common usage, the input to the operator is a grayscale image, as is the output. Pixel values at each point in the output represent the estimated absolute magnitude of the spatial gradient of the input image at that point [20].

2.2. Prewitt Edge Detector

The Prewitt edge detector is an appropriate way to estimate the magnitude and orientation of an edge. Although differential gradient edge detection needs a rather time consuming calculation to estimate the orientation from the magnitudes in the x and y-directions, the compass edge detection obtains the orientation directly from the kernel with the maximum response. The Prewitt operator is limited to 8 possible orientations, however experience shows that most direct orientation estimates are not much more accurate. This gradient based edge detector is estimated in the 3 × 3 neighbourhood for eight directions as shown in Figure 2. All the eight convolution masks are calculated. One convolution mask is then selected, namely that with the largest module [20].

1	0
0	−1

G_x

0	−1
1	0

G_y

Figure 1. Roberts gradient estimation operator.

−1	+1	+1
−1	−2	+1
−1	+1	+1

0°

+1	+1	+1
−1	−2	+1
−1	−1	+1

45°

Figure 2. Prewitt gradient estimation operator.

2.3. Sobel Edge Detector

The operator consists of a pair of 3 × 3 convolution kernels as shown in Figure 3. One kernel is simply the other rotated by 90°.

These kernels are designed to respond maximally to edges running vertically and horizontally relative to the pixel grid, one kernel for each of the two perpendicular orientations. The kernels can be applied separately to the input image, to produce separate measurements of the gradient component in each orientation (call these G_x and G_y). These can then be combined together to find the absolute magnitude of the gradient at each point and the orientation of that gradient [20]. The gradient magnitude is given by:

$$|G| = \sqrt{G_x^2 + G_y^2}$$

Typically, an approximate magnitude is computed using:

$$|G| = |G_x| + |G_y|$$

which is much faster to compute.

The angle of orientation of the edge (relative to the pixel grid) giving rise to the spatial gradient is given by:

$$\theta = \arctan\left(G_y / G_x\right)$$

2.4. Canny Edge Detector

This edge detector is due to J.F. Canny [19] (a recursive implementation of this algorithm was presented in [24]). In his work Canny specified three main criteria for the performance of edge detectors: First criteria, (low error rate) minimum number of false negatives and false positives. Second criteria, good localization, report edge location at correct position. In other words, the distance between the edge pixels as found by the detector and the actual edge is to be at a minimum. A third criterion is to have only one response to a single edge. In order to implement the canny edge detector algorithm, a series of steps must be followed.

−1	0	+1
−2	0	+2
−1	0	+1

G_x

+1	+2	+1
0	0	0
−1	−2	−1

G_y

Figure 3. Sobel gradient estimation operator.

Step 1: The first step is to filter out any noise in the original image before trying to locate and detect any edges. It uses a filter based on a Gaussian (bell curve), where the raw image isconvolved with a Gaussian filter. The result is a slightly blurred version of the original which is not affected by a single noisy pixel to any sig nificant degree. The Gaussian mask used in my implementation is shown in Figure 4 with σ = 1.4.

Step 2: After smoothing the image and eliminating the noise, the next step is to find the edge strength by taking the gradient of the image using the Sobel operator uses a pair of 3 × 3 convolution masks. The approximate gradient magnitude is given by $|G| = |G_x| + |G_y|$.

Step 3: Finding the edge direction is trivial once the gradient in the x and y directions are known. However, you will generate an error whenever sumX is equal to zero. So in the code there has to be a restriction set whenever this takes place. Whenever the gradient in the x direction is equal to zero, the edge direction has to be equal to 90 degrees or 0 degrees, depending on what the value of the gradient in the y-direction is equal to. If G_y has a value of zero, the edge direction will equal 0 degrees. Otherwise the edge direction will equal 90 degrees. The formula for finding the edge direction is just: $\theta = \arctan\left(G_y / G_x\right)$

Step 4: Once the edge direction is known, the next step is to relate the edge direction to a direction that can be traced in an image. So if the pixels of a 5 × 5 image are aligned as follows in Figure 5.

Then, it can be seen by looking at pixel "a", there are only four possible directions when describing the surrounding pixels, 0 degrees (in the horizontal direction), 45 degrees (along the positive diagonal), 90 degrees (in the vertical direction), or 135 degrees (along the negative diagonal). So now the edge orientation has to be resolved into one of these four directions depending on which direction it is closest to (e.g. if the orientation angle is found to be 3 degrees, make it zero degrees). Think of this as taking a semicircle and dividing it into 5 regions as shown in Figure 6.

Therefore, any edge direction falling within the range (0 to 22.5 & 157.5 to 180 degrees) is set to 0 degrees. Any edge direction falling in the range (22.5 to 67.5 degrees) is set to 45 degrees. Any edge direction falling in the range (67.5 to 112.5 degrees) is set to 90 degrees. And finally, any edge direction falling within the range (112.5 to 157.5 degrees) is set to 135 degrees.

$$\frac{1}{159}$$

2	4	5	4	2
4	9	12	9	4
5	12	15	12	5
4	9	12	9	4
2	5	5	5	2

Figure 4. Discrete approximation to Gaussian function with σ = 1.4.

Step 5: After the edge directions are known, nonmaximum suppression now has to be applied. Nonmaximum suppression is used to trace along the edge in the edge direction and suppress any pixel value (sets it equal to 0) that is not considered to be an edge. This will give a thin line in the output image.

X	X	X	X	X
X	X	X	X	X
X	X	*a*	X	X
X	X	X	X	X
X	X	X	X	X

Figure 5. The pixel "a" and the possible directions.

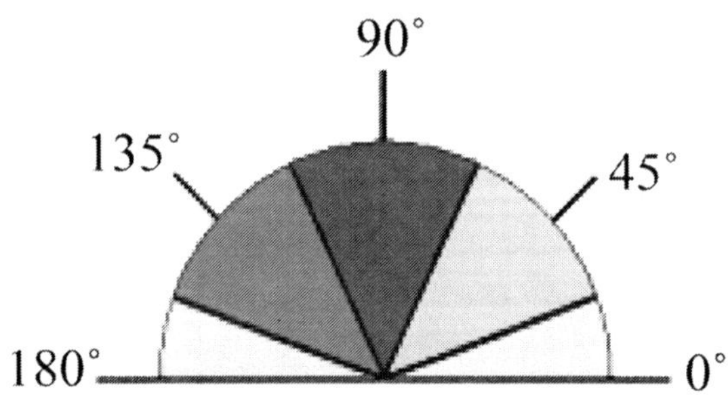

Figure 6. The range of edge direction in the five regions.

Step 6: Finally, hysteresis is used as a means of eliminating streaking. Streaking is the breaking up of an edge contour caused by the operator output fluctuating above and below the threshold. If a single threshold, T_1 is applied to an image, and an edge has an average strength equal to T_1, then due to noise, there will be instances where the edge dips below the threshold. Equally it will also extend above the threshold making an edge look like a dashed line. To avoid this, hysteresis uses 2 thresholds, a high and a low. Any pixel in the image that has a value greater than T_1 is presumed to be an edge pixel, and is marked as such immediately. Then, any pixels that are connected to this edge pixel and that have a value greater than T_2 are also selected as edge pixels. If you think of following an edge, you need a gradient of T_2 to start but you don't stop till you hit a gradient below T_1.

3. HAVRDA & CHARVAT'S ENTROPY

Regarding the statistical approach for describing texture, one of the simplest computational approaches is to use statistical moments of the gray level histogram of the image. The image histogram carries important information about the content of an image and can be used for discriminating the abnormal tissue from the local healthy background. Considering the gray level histogram $\{h_i, i = 0,1,2,\cdots,N_{g-1}\}$ where N_g is the number of distinct gray levels in the ROI (region of interest). If n is the total number of pixels in the region, then the normalized histogram of the ROI is the set $\{H_i, i = 0,1,2,\cdots,N_{g-1}\}$, where $H_i = h_i/n$. The source symbol probabilities is $H = \{H_i, i = 0,1,2,\cdots,N_{g-1}\}$. This set of probabilities must satisfy the condition, $0 \le H_i \le 1$. The average information per source output, denoted S(H) [25], Shannon entropy may be described as:

$$S(H) = -\sum_{i=0}^{N_{g-1}} H_i \log(H_i) \tag{2}$$

If we consider that a system can be decomposed in two statistical independent subsystems A and B, the Shannon entropy has the extensive property (additivity) $S(A+B) = S(A) + S(B)$, this formalism has been shown to be restricted to the Boltzmann-Gibbs-Shannon (BGS) statistics.

However, for non-extensive systems, some kind of extension appears to become necessary. Havrda & Charvat's [26,27] has proposed a generalization of the BGS statistics which is useful for describing the thermo statistical properties of non-extensive systems. It is based on a generalized entropic form,

$$HC_\alpha = \frac{1}{1-\alpha}\left(\sum_{i=0}^{N_g-1} H_i^\alpha - 1\right) \tag{3}$$

where the real number α is a entropic index that characterizes the degree of non-extensivity. This expression recovers to BGS entropy in the limit $\alpha \to 1$. Havrda & Charvat's entropy has a non-extensive property for statistical independent systems, defined by the following rule [28]:

$$HC_\alpha(A+B) = HC_\alpha(A) + HC_\alpha(B) + (1-\alpha) \cdot HC_\alpha(A) \cdot HC_\alpha(B). \tag{4}$$

Similarities between Boltzmann-Gibbs and Shannon entropy forms give a basis for possibility of generalization of the Shannon's entropy to the Information Theory. This generalization can be extended to image processing areas, specifically for the image segmentation, applying Havrda & Charvat's entropy to threshold images, which have non-additive information content.

Considering $HC_\alpha \geq 0$ in the pseudo-additive formalism of Equation (4), three different entropies can be defined with regard to different values of α.

For $\alpha < 1$, the Havrda & Charvat's entropy becomes a "sub extensive entropy" where:

$$HC_\alpha(A+B) < HC_\alpha(A) + HC_\alpha(B) \tag{5}$$

For $\alpha = 1$, the Havrda & Charvat's entropy reduces to an standard "extensive entropy" where:

$$HC_\alpha(A+B) = HC_\alpha(A) + HC_\alpha(B) \tag{6}$$

For $\alpha > 1$, the Havrda & Charvat's entropy becomes a "super extensive entropy" where:

$$HC_\alpha(A+B) > HC_\alpha(A) + HC_\alpha(B) \tag{7}$$

Let f(x, y) be the gray value of the pixel located at the point (x, y). In a digital image $\{f(x,y)|x \in \{1,2,\cdots,M\}, y \in \{1,2,\cdots,N\}\}$ of size M × N, let the histogram be h(a) for $a \in \{0,1,2,\cdots,255\}$ with f as the amplitude (brightness) of the image at the real coordinate position (x, y). For the sake of convenience, we denote the set of all gray levels $\{0,1,2,\cdots,255\}$ as G. Global threshold selection methods usually use the gray level histogram of the image. The optimal threshold t^* is determined by optimizing a suitable criterion function obtained from the gray level distribution of the image and some other features of the image.

Let t be a threshold value and B = $\{b_0, b_1\}$ be a pair of binary gray levels with $\{b_0, b_1\} \in G$. Typically b_0 and b_1 are taken to be 0 and 1, respectively. The result of thresholding an image function f(x, y) at gray level t is a binary function $f_t(x,y)$ such that $f_t(x,y) = b_0$ if $f_t(x,y) \leq t$ otherwise, $f_t(x,y) = b_1$. In general, a thresholding method determines the value $t^*$of t based on a certain criterion function. If t^* is determined solely from the gray level of each pixel, the thresholding method is point dependent [25].

Let $h_1, h_2, \cdots, h_k$ be the probability distribution for an image with k gray-levels. From this distribution, we derive two probability distributions, one for the object (class A) and the other for the background (class B), given by:

$$\lambda_A : \frac{h_1}{H_A}, \frac{h_2}{H_A}, \cdots, \frac{h_t}{H_A},$$
$$\lambda_B : \frac{h_{t+1}}{H_B}, \frac{h_{t+2}}{H_B}, \cdots, \frac{h_k}{H_B} \quad (8)$$

and where

$$H_A = \sum_{i=1}^{t} h_i, \quad H_B = \sum_{i=t+1}^{k} h_i \quad (9)$$

The Havrda & Charvat's entropy of order q for each distribution is defined as:

$$HC_\alpha(A) = \frac{1}{1-\alpha}\left(\sum_{i=1}^{t} \lambda_A^\alpha - 1\right)$$

$$\text{and } HC_\alpha(B) = \frac{1}{1-\alpha}\left(\sum_{i=t+1}^{k} \lambda_B^\alpha - 1\right) \tag{10}$$

The Havrda & Charvat's entropy HC_α is parametrically dependent upon the threshold value t for the foreground and background. It is formulated as the sum each entropy, allowing the pseudo-additive property, defined in Equation (3). We try to maximize the information measure between the two classes (object and background). When HC_α is maximized, the luminance level t that maximizes the function is considered to be the optimum threshold value.

$$t^*(\alpha) = \underset{t\in G}{\text{Argmax}}\left[HC_\alpha(A) + HC_\alpha(B) + (1-\alpha) \cdot HC_\alpha(A) \cdot HC_\alpha(B)\right]. \tag{11}$$

In the proposed scheme, first create a binary image by choosing a suitable threshold value using Havrda & Charvat's entropy. The technique consists of treating each pixel of the original image and creating a new image, such that $f_t(x,y) = 0$ if $f_t(x,y) \le t^*(q)$ otherwise, $f_t(x,y) = 1$ for every $x \in \{1,2,\cdots,M\}$, $y \in \{1,2,\cdots,N\}$.

When $\alpha \to 1$, the threshold value in Equation (3), equals to the same value found by Shannon's method. Thus this proposed method includes Shannon's method as a special case. The following expression can be used as a criterion function to obtain the optimal threshold at $\alpha \to 1$.

$$t^*(\alpha = 1) = \underset{t\in G}{\text{Argmax}}\left[HC_\alpha(A) + HC_\alpha(B)\right]. \tag{12}$$

The Havrda_Charvat_T procedure to select suitable threshold value t^* with α for grayscale image f can now be described as follows:

Procedure Havrda_Charvat_T,

Input: An image f of size r × c, and $\alpha > 0$.

Output: optimal threshold t^* of f.

Begin 1. Let f(x, y) be the original gray value of the pixel at the point (x, y), x = 1.. r, y = 1.. c.

2. Calculate the probability distribution $0 \le h_i \le 255$.

3. For all t $\in$ {0, 1, ..., 255}i. Calculate H_A, H_B, λ_A, and λ_B using Equations (8) and (9).

ii. Find optimum threshold value t*, such that

$$t^*(\alpha) = \underset{t \in G}{\text{Argmax}}\left[HC_\alpha(A) + HC_\alpha(B) + (1-\alpha) \cdot HC_\alpha(A) \cdot HC_\alpha(B)\right].$$

End.

The technique consists of treating each pixel of the original image and creating a new image, such that $f_t(x, y) = 0$ if $f_t(x, y) \le t^*(\alpha)$ otherwise, $f_t(x, y) = 1$ for every $x \in \{1, 2, \cdots, M\}$, $y \in \{1, 2, \cdots, N\}$.

4. DETECTING OF THE EDGES

We will use the usual masks for detecting the edges [29]. A spatial filter mask may be defined as a matrix w of size m × n. Assume that m = 2μ + 1 and n = 2ρ + 1, where μ, ρ are nonzero positive integers. For this purpose, smallest meaningful size of the mask is 3 × 3. Such mask coefficients, showing coordinate arrangement as Figure 7(a). Image region under the above mask is shown asFigure 7(b).

In order to edge detection, firstly classification of all pixels that satisfy the criterion of homogeneousness, and detection of all pixels on the borders between different homogeneous areas. In the proposed scheme, first create a binary image by choosing a suitable threshold value using Havrda & Charvat entropy. Window is applied on the binary image. Set all window coefficients equal to 1 except centre, centre equal to × as shown in Figure 8.

Move the window on the whole binary image and find the probability of each central pixel of image under the window. Then, the entropy of each central pixel of image under the window is calculated as S(CPix) = $-p_c \ln(p_c)$. Where, p_c is the probability of central pixel CPix of binary image under the window. When the probability of central pixel, $p_c = 1$, then the entropy of this pixel is zero. Thus, if the gray level of all pixels under the window homogeneous, $p_c = 1$ and S = 0. In this case, the central pixel is not an edge pixel. Other possibilities of entropy of central pixel under window are shown in Table 1.

$w(-1,-1)$	$w(-1,0)$	$w(-1,1)$
$w(0,-1)$	$w(0,0)$	$w(0,1)$
$w(1,-1)$	$w(1,0)$	$w(1,1)$

(a)

$f(x-1,y-1)$	$f(x-1,y)$	$f(x-1,y+1)$
$f(x,y-1)$	$f(x,y)$	$f(x,y+1)$
$f(x+1,y-1)$	$f(x+1,y)$	$f(x+1,y+1)$

(b)

Figure 7. Coordinate arrangement and image region under 3 × 3 mask.

In cases $p_c = 8/9$, and $p_c = 7/9$, the diversity for gray level of pixels under the window is low. So, in these cases, central pixel is not an edge pixel. In remaining cases, $p_c \leq 6/9$, the diversity for gray level of pixels under the window is high. The complete algorithm can now be described as follows:

1	1	1
1	×	1
1	1	1

Figure 8. Window coefficients of 3 × 3 mask.

Table 1. p and S of central under window.

p	1/9	2/9	3/9	4/9
S	0.2441	0.3342	0.3662	0.3604
p	5/9	6/9	7/9	8/9
S	0.3265	0.2703	0.1955	0.1047

Algorithm HCEdgeDetection;

Input: A grayscale image A (M × N).

Output: The edge detection image g.

Begin

1. Select suitable t^*, α, using Havrda_Charvat_T procedure.

2. Create a binary image:

If $f(x, y) \leq t^* \alpha)$ then $f(x,y) = 0$ Else $f(x, y) = 1$.

3. Create a mask w, with dimensions m × n:

Normally, $m = n = 3$. $\mu = (m-1)/2$ and $\rho = (n-1)/2$.

4. For all $1 \leq x \leq M$ and $1 \leq y \leq N$:

Find g an output image by set g = f.

5. For all $\rho + 1 \leq y \leq N-\rho$ and $\mu + 1 \leq x \leq M-\mu$checking for edge pixels:

i. sum = 0;

ii. For all $-\rho \leq k \leq \rho$ and $-\mu \leq j \leq \mu$:

If (f(x, y) = f (x+j, y+k)) Then Sum = sum + 1.

iii. If (sum > 6) Then g(x,y)=0 Else g(x,y) = 1.

End algorithm.

5. PROPOSED ALGORITHM

Here, the algorithm produces three different threshold values t_1, t_2 and t_3. We use Havrda_Charvat_T procedure, to find the threshold value t_1 through the entire image. Then we split the image by t_1 into two grayscale parts, the object and background. Applying the Equation (11), to find the locals threshold values t_2 and t_3 of object and background, respectively. Independently, we apply HC EdgeDetection Procedure with threshold values t_1, t_2 and t_3. We merge the resultant edge images to obtain the reconstructed edge image.

In order to minimize the execution time, we deal with the histogram vectors, $0, 1, \cdots, t_1$ and $t_1 + 1, \cdots, 255$ of object and background parts, respectively rather than the matrices size of them.

The steps of proposed algorithm are as follows:

1- Read the grayscale level image, I = imread ('test. tif');

2- Calculate the histogram H with 256 elements of I;

3- Call Havrda_Charvat_T(H, α) to find the optimal threshold value, t_1;

4- Divide H into two parts H_{Low} and H_{High} using t_1.

5- Recall Havrda_Charvat_T(H_{Low}, α) to find the optimal threshold value t_2;

6- Recall Havrda_Charvat_T(H_{High}, α) to find the optimal threshold value t_3;

7- Now we have 3 values of the threshold, $t_2 < t_1 < t_3$. Reconstruct bitmap image f, such that: if ($I_{x,y} < t_1$ and $I_{x,y} >= t_2$) or ($I_{x,y} >= t_3$) then $f_{x,y} = 1$; end;

8- Call the procedure HCEdgeDetection(f) to find edge image g.

9- Display the image g, imshow(g);

The above procedures can be done together in the following MATLAB program:

```
I=imread('test.tif');
alpha=0.1;
[M,N,R]=size(I);
if R==3 I=rgb2gray(I); end;
H = imhist(I);
[T1, Max1, Loc1] =
                    Havrda_Charvat_T(H,alpha);
HLow= H (1:Loc1-1,:);
[T2, Max2, Loc2]=
                 Havrda_Charvat_T(HLow,alpha);
HHigh= H (Loc1:size(H),:);
[T3, Max3, Loc3] =
                 Havrda_Charvat_T(HHigh,alpha);
f=zeros(M,N);
for i=1:M;
   for j=1:N;
   if ((I(i,j) >= T2)&(I(i,j) < T1))
        |(I(i,j) >= T3) f(i,j)=1; end;
   end;
end
[g]= HCEdgeDetection(f);
figure; imshow(g);
```

6. RESULTS AND DISCUSSION

In order to test the method proposed in this paper and compare with the other edge detectors, common gray level test images with different resolutions and sizes are detected by Canny, LOG, Roberts, Prewitt, Sobel and the proposed method respectively. The performance of the proposed scheme is evaluated through the simulation results using MATLAB. Prior to the application of this algorithm, no pre-processing was done on the tested images (Figure 9).

The algorithm has two main phases global and local enhancement phase of the threshold values and detection phase, we present the results of implementation on these images separately. Here, we have used in addition to the original gray level function f(x, y), a function g(x, y) that is the average gray level value in a 3 × 3 neighborhood around the pixel (x, y). We use MATLAB to calculate the average time for each method at different images size by repeating 10 times for each type of image. As shown in Figure 10, the chart of the test images and the average of run time for the classical methods and proposed scheme.

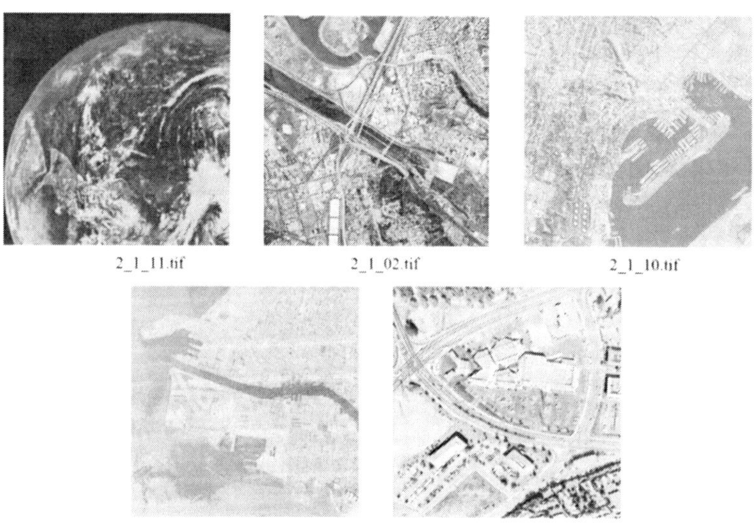

Figure 9. Samples of test images.

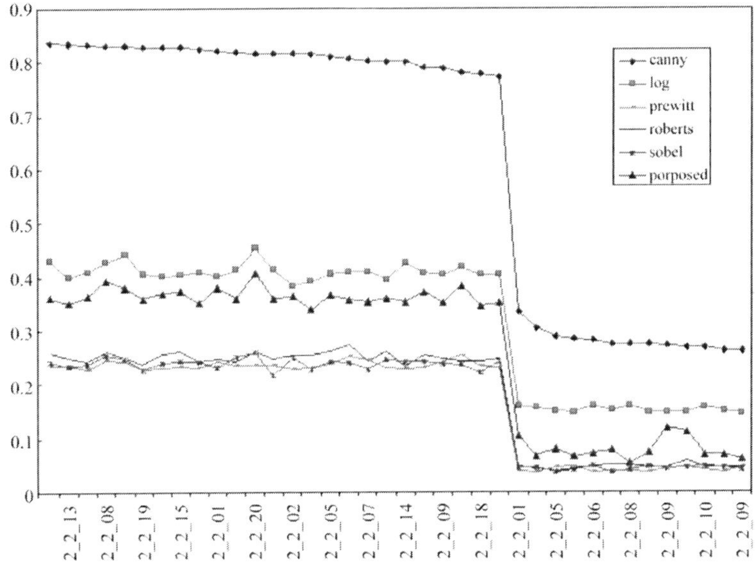

Figure 10. Comparison of run time between some classical methods and proposed method on the same datasets.

It has been observed that the proposed edge detector works effectively for different gray scale digital images as compare to the run time of classical methods. Some selected results of edge detections for these test images using the classical methods and proposed scheme are shown inFigure 11 and Table 2.

10. CONCLUSION

This paper shows the new algorithm based on the Havrda & Charvat's entropy for edge detection using split and merge technique of the histogram of grey scale image. The objective is to find the best edge representation and minimize the computation time. A set of experiments in the domain of edge detection are presented on a sample of test images, see Figure 9. An edge detection performance is compared to the previous classic methods, such as Canny, LOG, and Sobel. Analysis shows that the effect of the proposed method is better than those methods in execution time, and it is also considered as easy implementation. The significance of this study lies in decreasing the computation time with generate suitable quality of edge detection. It is already pointed out in the introduction that the traditional methods give rise to the exponential increment of computational time. Experiment results have demonstrated that the proposed scheme for edge detection can be used for different gray level digital images. Another benefit comes from easy implementation of this method. The proposed method works well as compared to the previous methods, Roberts, Prewitt, and Sobel on test samples.

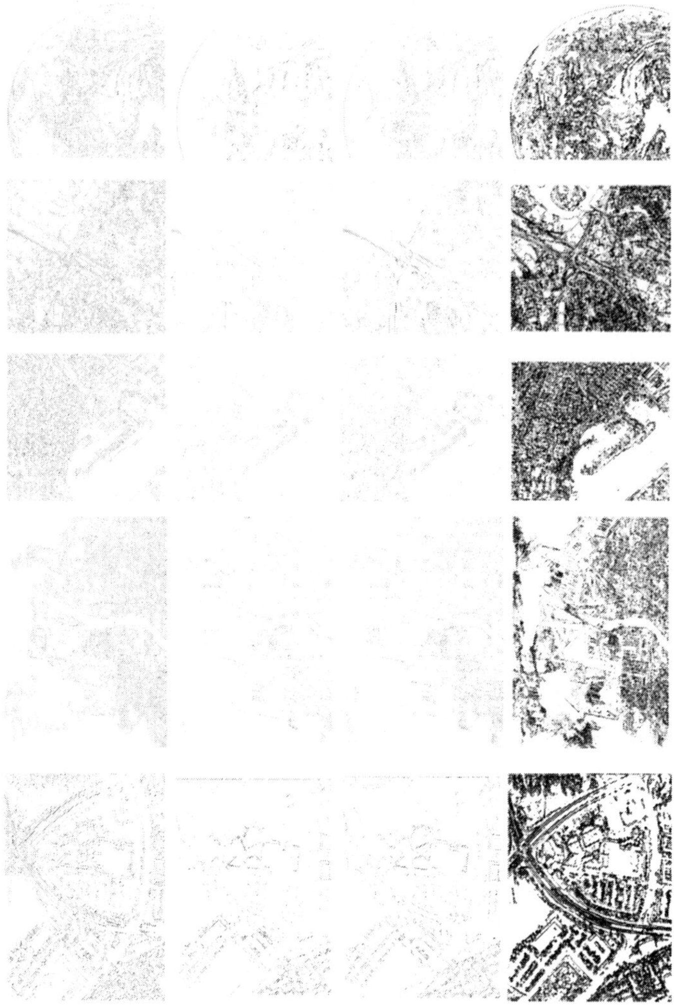

Figure 11. Edge detections of test images using the LoG method, Roberts method, Sobel method and proposed method, respectively.

Table 2. The threshold values of tested images with different values of α.

Image Name	$\alpha = 0.1$			$\alpha = 0.5$			$\alpha = 0.7$			$\alpha = 0.9$			$\alpha = 1.2$		
	t_1	t_2	t_3	t_1	t_2	t_3	t_1	t_2	t_3	t_1	t_2	t_3	t_1	t_2	t_3
2_1_01	142	69	182	139	78	178	139	80	178	139	83	176	227	226	227
2_1_02	96	53	164	102	61	164	104	64	164	136	87	180	231	230	231
2_1_03	142	107	196	142	109	195	142	108	195	142	108	173	221	220	221
2_1_04	176	105	196	123	94	174	124	95	172	124	95	171	211	210	211
2_1_05	137	66	184	136	66	180	136	66	179	136	66	178	234	233	234
2_1_06	62	52	146	143	52	188	144	61	188	144	105	187	236	235	236
2_1_07	146	111	202	150	117	197	151	126	191	151	126	186	230	229	230
2_1_08	158	132	207	161	136	193	161	136	191	161	141	189	227	226	227
2_1_09	140	120	176	174	140	201	175	140	201	175	140	200	231	230	231
2_1_10	130	114	182	173	130	201	173	130	199	174	130	199	230	229	230
2_1_11	130	92	168	130	94	165	130	94	165	130	94	164	208	207	208
2_1_12	152	128	189	166	141	193	169	144	194	170	145	193	224	223	224
2_2_01	124	69	190	121	71	176	120	71	173	119	70	170	234	233	234
2_2_02	132	120	171	171	130	198	171	130	197	171	129	196	229	228	229
2_2_03	140	122	171	167	138	194	168	138	193	168	138	192	223	222	223
2_2_04	115	95	210	161	118	211	162	122	209	163	129	208	235	234	235
2_2_05	143	69	184	142	86	181	142	90	181	142	92	180	232	231	232
2_2_06	151	114	189	152	127	188	152	127	188	151	127	188	222	221	222
2_2_07	146	120	213	149	118	182	150	118	181	166	144	189	226	225	226
2_2_08	126	72	192	123	74	178	122	74	175	121	73	171	235	234	235

2_2_09	104	68	141	104	69	143	104	69	144	144	104	181	231	230	231
2_2_10	108	86	211	109	87	152	109	87	153	108	88	154	231	230	231
2_2_11	99	85	194	98	70	194	98	70	194	98	45	195	233	232	233
2_2_12	80	62	188	82	65	168	83	65	167	83	65	166	254	249	254
2_2_13	162	115	189	165	121	189	166	119	189	168	119	191	226	225	226
2_2_14	62	44	180	169	60	200	154	98	191	152	99	189	235	231	235
2_2_15	118	88	189	124	105	173	159	121	195	159	121	194	223	222	223
2_2_16	156	78	194	154	95	190	153	103	189	153	105	187	228	224	228
2_2_17	71	57	144	70	57	141	68	45	141	141	68	173	219	218	219
2_2_18	97	80	192	193	126	212	194	130	212	194	131	212	234	233	234
2_2_19	111	87	193	110	86	191	191	110	209	191	139	207	232	231	232
2_2_20	85	76	169	166	79	193	165	79	192	165	79	192	233	232	233
2_2_21	99	47	208	99	46	162	101	52	162	101	47	161	237	236	237
2_2_22	126	103	211	147	116	212	149	121	200	151	122	181	233	232	233
2_2_23	128	29	170	129	29	170	129	29	170	129	28	170	229	228	229
2_2_24	171	53	200	173	127	198	173	127	198	174	127	198	234	233	234
3_2_25	200	115	224	162	115	208	160	115	208	160	115	208	245	242	245
5_2_09	129	71	218	129	75	177	129	75	177	129	75	176	255	252	255

REFERENCES

1. S. S. Alamri, N. V. Kalyankar and S. D. Khamitkar, "Image Segmentation By Using Edge Detection," International Journal on Computer Science and Engineering (IJCSE), Vol. 2, No. 3, 2010, pp. 804-807.

2. J. Lzaro, J. L. Martn, J. Arias, A. Astarloa and C. Cuadrado, "Neuro Semantic Thresholding Using OCR Software for High Precision OCR Applications," Image Vision Computing, Vol. 28, No. 4, 2010, pp. 571-578.

3. Z. Xue, D. Ming, W. Song, B. Wan and S. Jin, "Infrared Gait Recognition Based on Wavelet Transform and Support Vector Machine," Pattern Recognition, Vol. 43, No. 8, 2010, pp. 2904-2910.

4. M. A. El-Sayed, "Edges Detection Based on Renyi Entropy with Split/Merge," Computer Engineering and Intelligent Systems (CEIS), Vol. 3, No. 9, 2012, pp. 32-41.

5. G. C. Anagnostopoulos, "SVM-Based Target Recognition from Synthetic Aperture Radar Images Using Target Region Outline Descriptors," Nonlinear Analysis: Theory, Methods & Applications, Vol. 71, No. 12, 2009, pp. e2934-e2939.

6. Y.-T. Hsiao, C.-L. Chuang, Y.-L. Lu and J.-A. Jiang, "Robust Multiple Objects Tracking Using Image Segmentation and Trajectory Estimation Scheme in Video Frames," Image Vision Computing, Vol. 24, No. 10, 2006, pp. 1123-1136.

7. M. T. Doelken, H. Stefan, E. Pauli, A. Stadlbauer, T. Struffert, T. Engelhorn, G. Richter, O. Ganslandt, A. Doerfler and T. Hammen, "1H-MRS Profile in MRI Positive versus MRI Negative Patients with Temporal Lobe Epilepsy," Seizure, Vol. 17, No. 6, 2008, pp. 490-497.

8. S. Kresic-Juric, D. Madej and S. Fadil, "Applications of Hidden Markov Models in Bar Code Decoding," International Journal of Pattern Recognition Letters, Vol. 27, No. 14, 2006, pp. 1665-1672.

9. G. Markus, Essam A. El-Kwae and R. K. Mansur, "Edge Detection in Medical Images Using a Genetic Algorithm," IEEE Transactions on Medical Imaging, Vol. 17, No. 3, 1998, pp. 469-474.

10. M. Wang and Y. Shuyuan, "A Hybrid Genetic Algorithm Based Edge Detection Method for SAR Image," IEEE Proceedings of the Radar Conference'05, 9-12 May 2005, pp. 1503-506.

11. A. El-Zaart, "A Novel Method for Edge Detection Using 2 Dimensional Gamma Distribution," Journal of Computing Science, Vol. 6, No. 2, 2010, pp. 199-204.

12. M. A. El-Sayed, "A New Algorithm Based Entropic Threshold for Edge Detection in Images," International Journal of Computer Science Issues (IJCSI), Vol. 8, No. 5, 2011, pp. 71-78.

13. V. Aurich and J. Weule, "Nonlinear Gaussian Filters Performing Edge Preserving Diffusion," Proceeding of the 17th Deutsche Arbeitsgemeinschaft für Mustererkennung (DAGM) Symposium, Bielefeld, 13-15 September 1995, Springer-Verlag, pp. 538-545.

14. M. Basu, "A Gaussian Derivative Model for Edge Enhancement," Pattern Recognition, Vol. 27, No. 11, 1994, pp. 1451-1461.

15. G. Deng and L. W. Cahill, "An Adaptive Gaussian Filter for Noise Reduction and Edge Detection," Proceeding of the IEEE Nuclear Science Symposium and Medical Imaging Conference, San Francisco, 31 October-6 November 1993, IEEE Xplore Press, pp. 1615-1619.

16. C. Kang and W. Wang, "A Novel Edge Detection Method Based on the Maximizing Objective Function," Pattern Recognition, Vol. 40, No. 2, 2007, pp. 609-618.

17. Q. Zhu, "Efficient Evaluations of Edge Connectivity and Width Uniformity," Image Vision Computing, Vol. 14, No. 1, 1996, pp. 21-34.

18. B. Mitra, "Gaussian Based Edge Detection Methods—A Survey," IEEE Transactions on Systems, Man and Cybernetics, Vol. 32, No. 3, 2002, pp. 252-260.

19. J. F. Canny, "A Computational Approach to Edge Detection," IEEE Transactions on Pattern Analysis and Machine Intelligence (IEEE TPAMI), Vol. 8, No. 6, 1986, pp. 769-798.

20. N. Senthilkumaran and R. Rajesh, "Edge Detection Techniques for Image Segmentation—A Survey," Proceedings of the International

Conference on Managing Next Generation Software Applications (MNGSA-08), 2008, pp. 749-760.

21. K. W. Bowyer, C. Kranenburg and S. Dougherty, "Edge Detector Evaluation Using Empirical ROC Curves," IEEE Conference on Computer Vision and Pattern Recognition (CVPR), 1999, pp. 354-359.

22. M. Heath, S. Sarkar, T. Sanocki and K. W. Bowyer, "A Robust Visual Method for Assessing the Relative Performance of Edge Detection Algorithms," IEEE Transactions on Pattern Analysis and Machine Intelligence (TPAMI), Vol. 19, No. 12, 1997, pp. 1338-1359.

23. M. Shin, D. Goldgof and K. W. Bowyer, "An Objective Comparison Methodology of Edge Detection Algorithms for Structure from Motion Task," IEEE Conference on Computer Vision and Pattern Recognition (CVPR), 1998, pp. 190-195.

24. R. Deriche, "Using Canny's Criteria to Derive a Recursively Implemented Optimal Edge Detector," International Journal of Computer Vision (IJCV), Vol. 1, No. 2, 1987, pp. 167-187.

25. C. E. Shannon, "A Mathematical Theory of Communication," Bell System Technical Journal, Vol. 27, No. 3, 1948, pp. 379-423.

26. A. P. S. Pharwaha and B. Singh, "Shannon and NonShannon Measures of Entropy for Statistical Texture Feature Extraction in Digitized Mammograms," Proceedings of the World Congress on Engineering and Computer Science (WCECS), San Francisco, Vol. II, 20-22 October 2009.

27. C. Tsallis, "Possible Generalization of Boltzmann-Gibbs Statistics," Journal of Statistical Physics, Vol. 52, No. 1-2, 1988, pp. 479-487.

28. M. P. de Albuquerque, I. A. Esquef and A. R. Gesualdi Mello, "Image Thresholding Using Tsallis Entropy," Pattern Recognition Letters, Vol. 25, No. 9, 2004, pp. 1059- 1065.

29. B. Singh and A. P. Singh, "Edge Detection in Gray Level Images Based on the Shannon Entropy," Journal of Computing Science, Vol. 4, No. 3, 2008, pp. 186-191.

Chapter 10

Edge Detection with a Preprocessing Approach

Mohamed Abo-Zahhad[1], Reda Ragab Gharieb[1], Sabah M. Ahmed[1], Ahmed Abd El-Baset Donkol[2]

[1]Department of Electrical Engineering Communication and Electronics, Assuit University, Assiut, Egypt

[2]Department of Electrical Engineering and Computer Sciences, Nahda University, Beni-Suef, Egypt

ABSTRACT

Edge detection is the process of determining where boundaries of objects fall within an image. So far, several standard operators-based methods have been widely used for edge detection. However, due to inherent quality of images, these methods prove ineffective if they are applied without any preprocessing. In this paper, an image preprocessing approach has been adopted in order to get certain parameters that are useful to perform better edge detection with the standard operators-based edge detection methods. The proposed preprocessing approach involves computation of the histogram, finding out the total number of peaks and suppressing irrelevant peaks. From the intensity values corresponding to relevant peaks, threshold values are obtained. From these threshold values, optimal multilevel thresholds are calculated using the Otsu method, then multilevel image segmentation is carried out. Finally, a standard edge detection method can be applied to the resultant segmented image.

Simulation results are presented to show that our preprocessed approach when used with a standard edge detection method enhances its performance. It has been also shown that applying wavelet edge detection method to the segmented images, generated through our preprocessing approach, yields the superior performance among other standard edge detection methods.

KEYWORDS

Discrete Wavelet Transform, Image Edge Detection, Robert, Prewitt, Sobel, Canny Techniques

1. INTRODUCTION

Edge detection is basically an image segmentation technique that divides spatial domain, on which the image is defined into meaningful parts or regions [1]. Edges characterize boundaries of objects in an image and object recognition is of fundamental importance in image processing. Edges typically occur on the boundary between two different regions in an image, where there is a more or less abrupt change in gray level or texture indicating the end of one region and the beginning of another. Edge detection allows user to observe those boundaries. Edge detection therefore finds practical applications in medical imaging, computer-guided surgery, locate object in satellite images, face recognition, and finger print recognition, automatic traffic controlling systems, study of anatomical structure etc. Several edge detection techniques have been developed such as gradient based standard operators like Robert, Prewitt, Sobel. However, they have not given sharp edges and have been highly sensitive and provided poor performance in noisy images [1] . Laplacian based Marr Hildrith operators also suffers from two limitations: high probability of detecting false edges and the localization error may be severe at curved edges but the algorithm proposed by Canny in 1986 is considered as the ideal edge detection algorithm for images that are corrupted with noise [2] . Canny's aim is to discover the optimal edge detection algorithm which reduces the probability of detecting false edges, and gives sharp edges.

Wavelet analysis is an efficient tool for edge detection. Deterministic methods are inappropriate when noise models are applied. In images with low SNR, the noise must be accounted with a noise model, multiplicative or additive [3] . An important issue in edge detection is the scale of detection filter. Small-scaled filters are sensitive to edge signals but also proneto noise, whereas large-scaled filters are robust to noise but could filter out fine details. As suggested by Marr and Hildreth [4] , multiple scales could be employed to describe and synthesize the varieties of edge structures. The idea of scale multiplication was first exploited in [5] , where it was shown that the scale products could improve the edge localization. Mallat [6] illustrated mathematically that signals and noise had different singularities and edge structures present observable magnitudes along the scales, while noise decreased rapidly. With this observation, Xu et al. [7] proposed a wavelet-based spatially selective filtering technique by multiplying the adjacent scales. Sadler and Swami [8] applied the wavelet-multiscale-products to step detection and estimation and Bao and Zhang [9] presented a denoising scheme by thresholding the multiscale products.

This paper presents a preprocessing approach to enhance the performance of standard operators-base edge detection methods. The approach consists of computation of the image histogram, finding out the total number of peaks and suppressing irrelevant peaks. From the intensity values corresponding to relevant peaks, threshold values are obtained. From these threshold values, optimal multilevel thresholds are calculated using the Otsu method, then multilevel image segmentation is carried out. Finally, a standard edge detection method is applied to the resultant segmented image. The rest of the paper is organized as follows: In Section 2, a review of standard operator-based edge detection methods are described, advantage and disadvantage of these methods are summarized and standard wavelet edge detection method is also explained. In Section 3, the edge detection with the proposed preprocessing approach is introduced. Section 4 presents simulation results and shows a quantitative comparison between several preprocessed standard edge detection methods and these standard methods without the preprocessing. We draw the conclusion in Section 5.

2. PREVIOUS WORK

Edge detection is the process of determining where edges of objects fall within an image. Figure 1shows schematic diagram for the standard edge detection method. To sense and detect abrupt change at edges, several operators have been constructed based on different ideas. In the following section, a brief review of some operators is presented.

2.1. First Order Edge Detector

The gradient method detects the edges by looking for the maximum and minimum in the first derivative of the image. Thus, consider the two dimensional function $f(x,y)$ to represent the input image then image gradient is given by the following formula [1] :

$$\nabla f = G[f(x,y)] = \begin{bmatrix} f_x \\ f_y \end{bmatrix} = \begin{bmatrix} \frac{\partial f(x,y)}{\partial x} \\ \frac{\partial f(x,y)}{\partial y} \end{bmatrix} \tag{1}$$

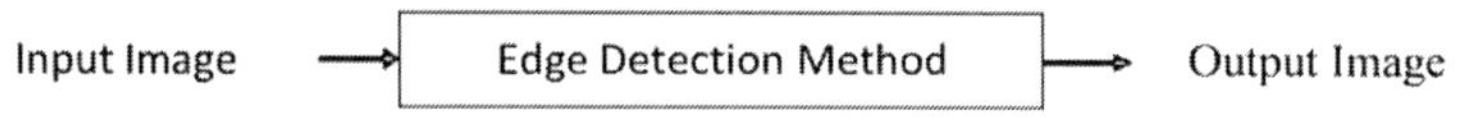

Figure 1. Standard edge detector.

The gradient magnitude can be computed by the formula:

$$|\nabla f| = \sqrt{f_x^2 + f_y^2} = \sqrt{\left(\frac{\partial f}{\partial x}\right)^2 + \left(\frac{\partial f}{\partial y}\right)^2} \approx |f_x| + |f_y| \tag{2}$$

The gradient direction can be computed by the formula:

$$\theta = \tan^{-1}(f_y / f_x) \tag{3}$$

The magnitude of the gradient computed by Equation (2) gives edge strength and the gradient direction is always perpendicular to the direction of the edge. Robert, Sobel, and Prewitt operators are classified as standardfirst order derivative operators which are easy to operate but highly sensitive to noise [1] .

2.1.1. Robert Operator

The input image is convolved with the default kernels of operator and gradient magnitude and directions are computed. It uses the following 2 × 2 two kernels [10] :

$$G_x = \begin{bmatrix} +1 & 0 \\ 0 & -1 \end{bmatrix} \text{ and } G_y = \begin{bmatrix} 0 & +1 \\ -1 & 0 \end{bmatrix} \quad (4)$$

The plus factor of this operator is its simplicity but having small kernel it is highly sensitive to noiseand not compatible with today's technology [10] .

2.1.2. Sobel Operator

It convolves the input image with kernel and computes the gradient magnitude and direction. It uses the following 3 × 3 two kernels:

$$G_x = \begin{bmatrix} -1 & 0 & +1 \\ -1 & 0 & +1 \\ -1 & 0 & +1 \end{bmatrix} \quad (5)$$

As compared to Robert operator, Sobel operator has slow computation. However it is less sensitive to noise as compared to Robert operator.

2.1.3. Prewitt Operator

Prewitt edge operator gives better performance than that of Sobel operator. The function of Prewitt edge detector is almost the same as Sobel detector but have different kernels:

$$G_x = \begin{bmatrix} -1 & 0 & +1 \\ -1 & 0 & +1 \\ -1 & 0 & +1 \end{bmatrix} \quad (6)$$

2.2. Second Order Edge Detector

Any definition of a second derivate 1) must be zero in flat areas; 2) must be nonzero at the onset and end of a gray level step and ramp; and 3) must be zero along ramps of constant slope [11] . The Laplacian operator ∇^2 for a 2D image f(x, y) is defined by the following formula [10] :

$$\nabla^2 f(x,y) = \frac{\partial^2}{\partial x^2} f(x,y) + \frac{\partial^2}{\partial y^2} f(x,y) \tag{7}$$

There are several ways to define digital Laplacian using neighborhoods. Whatever the definition, it has to satisfy the properties of a second derivate outlined in Equation (11). The following notation is used for second order derivative in the x and y directions [11]

$$\frac{\partial^2 f}{\partial^2 x^2} = f(x+1,y) + f(x-1,y) - 2f(x,y) \tag{8}$$

$$\frac{\partial^2 f}{\partial^2 y^2} = f(x,y+1) + f(x,y-1) - 2f(x,y) \tag{9}$$

$$\nabla^2 f = [f(x+1,y) + f(x-1,y) + f(x,y+1) + f(x,y-1)] - 4f(x,y) \tag{10}$$

Furthermore, when the first derivative is at a maximum, the second derivative is zero [12] .

2.2.1. Laplacian

Laplacian of an image $f(x,y)$ is defined by:

$$L[f(x,y)] = \nabla^2 f = [f(x+1,y) + f(x-1,y) + f(x,y+1) + f(x,y-1)] - 4f(x,y) \tag{11}$$

In this case, Laplacian kernel given by the following equation is adopted.

$$G_{xy} = \begin{bmatrix} 0 & 1 & 0 \\ 1 & -4 & 1 \\ 0 & 1 & 0 \end{bmatrix} \tag{12}$$

2.2.2. Laplacian of Gaussian (Marr-Hildreth Edge Detector)

The Laplacian of Gaussian (LoG) follows the following four steps [12] :

1) Smooth the image using Gaussian filter which is defined by:

$$G(x,y) = \frac{1}{\sqrt{2\pi\sigma^2}} e^{-\left(\frac{x^2+y^2}{2\sigma^2}\right)} \tag{13}$$

2) Enhance the edges using Laplacian operator which given by this equation [12] :

$$\mathrm{LoG}(x,y) = -\frac{1}{\pi\sigma^4}\left[1 - \frac{x^2+y^2}{2\sigma^2}\right] e^{-\frac{x^2+y^2}{2\sigma^2}} \tag{14}$$

3) Estimate the zero crossings denote the edge location.

4) Use linear interpolation to determine the sub-pixel location of the edge.

Greater the value of σ, broader the Gaussian filter, more is the smoothing. Too much smoothing may make the detection of edges difficult; also called the Maxican Hat operator.

2.2.3. Difference of Gaussian

To reduce the computational requirements, the LoG is approximated by the DoG. The width of the edge can be adjusted by changing σ_1 and σ_2. The difference of Gaussian (DoG) operator of an image $f(x,y)$ is defined by:

$$\mathrm{DoG}(x,y)=\frac{e^{-\left(\frac{x^2+y^2}{2\pi\sigma_1^2}\right)}}{2\pi\sigma_1^2}-\frac{e^{-\left(\frac{x^2+y^2}{2\pi\sigma_2^2}\right)}}{2\pi\sigma_2^2} \tag{15}$$

LoG requires large computation time for a large edge detector mask. Second Order Derivative Methods Properties. The following are the three most important properties of the second order derivatives:

1) Second order derivative methods especially Laplacian, are very sensitive to noise.

2) Probability of false and missing edges remains.

3) Localization is better than gradient operators.

2.3. Canny Edge Detector

The Canny edge detection algorithm is known to many as the optimal edge detector. Canny's intentions were to enhance the many edge detectors already out at the time he started his work. He was very successful in achieving his goal and his ideas and methods can be found [2] . In his paper, he followed a list of criteria to improve current methods of edge detection. The first and most obvious is low error rate. It is important that edges occurring in images should not be missed and that there be no responses to non-edges. The second criterion is that the edge points be well localized. In other words, the distance between the edge pixels as found by the detector and the actual edge is to be at a minimum. A third criterion is to have only one response to a single edge. This was implemented because the first two were not substantial enough to completely eliminate the possibility of multiple responses to an edge.

Based on these criteria, the Canny edge detector first smoothies the image to eliminate the noise. It then finds the image gradient to highlight regions with high spatial derivatives. The algorithm then tracks along these regions and suppresses any pixel that is not at the maximum (non-maximum suppression). The gradient array is now further reduced by hysteresis. Hysteresis is used to track along the remaining pixels that have not been suppressed. Hysteresis uses two thresholds and if the magnitude is below the first threshold T_1, it is set to zero (made a non-edge). If the magnitude is above the high threshold T_2, it is made an edge. And if the magnitude is between the two thresholds, then it is set to zero unless there is a path from this pixel to a pixel with a gradient above the second threshold T_2. In order to implement the canny edge detector algorithm, a series of steps must be followed [13] .

Step 1: Smooth the image with a two dimensional Gaussian filters.

Step 2: Take the gradient of the image.

Step 3: The direction of the edge is computed using the gradient in the x and y directions.

Step 4: relate the edge direction to a direction that can be traced in an image.

Step 5: Non-maximum suppression.

Step 6: Hysteresis.

2.4. Advantages and Disadvantages of Standard Edge Detection Methods

Each edge detection method has its advantages and disadvantages. Table 1 summarize the main advantages and disadvantages of each method [12] .

2.5. Edge Detection Using Wavelets

2.5.1. 2D Discrete Wavelet Transform Edge Detection

First we define a Gaussian equation in two dimensions as [14] [15]

$$\theta(x,y) = \frac{1}{2\pi\sigma_x\sigma_y} e^{\frac{1}{2}\left(\left(\frac{x-\mu_x}{\sigma_x}\right)^2 + \left(\frac{y-\mu_y}{\sigma_y}\right)^2\right)} \tag{16}$$

Table 1. Advantages and disadvantages of edge detector ion methods.

Method	Advantages	Disadvantages
Standard (Sobel, Prewitt, Kirsch)	Simplicity, detection of edges and their orientations	Sensitivity to noise, inaccurate
Zero crossing (Laplacian, 2nd directional derivative)	Detection of edges and their orientations. Having fixed characteristics in all directions	Responding to some of the existing edges, sensitivity to noise
Laplacian of Gaussian (LoG)-(Marr-Hildreth)	Finding the correct places of edges, testing wider area around the pixel	Malfunctioning at the corners, curves and where the gray level intensity function varies. Not finding the orientation of edge because of using the Laplacian filter
Gaussian (Canny, Shen-Castan)	Using probability for finding error rate, localization of response. Improving signal to noise ratio, better detection specially in noise conditions	Complex computations, false zero crossing, time consuming

where, σ is standard deviation of the signal, and we assume that $\mu_x = \mu_y = 0$. We define two wavelet functions $\Psi^1(x,y)$ and $\Psi_s^2(x,y)$ such that

$$\Psi^1(x,y) = \frac{\partial\theta(x,y)}{\partial x} \text{ and } \Psi^2(x,y) = \frac{\partial\theta(x,y)}{\partial y} \tag{17}$$

$$\Psi_s^1(x,y) = \frac{1}{s^2}\Psi_1\left(\frac{x}{s},\frac{y}{s}\right) \tag{18}$$

$$\Psi_s^2(x,y) = \frac{1}{s^2}\Psi_2\left(\frac{x}{s},\frac{y}{s}\right) \tag{19}$$

The wavelet transform of f(x, y) at the scale s has two components defined by

$$W_s^1 f(x,y) = f * \Psi_s^1(x,y) \text{ and } W_s^2 f(x,y) = f * \Psi_s^2(x,y) \tag{20}$$

$$\begin{pmatrix} w_s^1 f(x,y) \\ W_s^2 f(x,y) \end{pmatrix} = s\begin{pmatrix} \frac{\partial}{\partial x}(f*\theta_s)(x,y) \\ \frac{\partial}{\partial y}(f*\theta_s)(x,y) \end{pmatrix} = s\nabla(f*\theta_s)(x,y) \tag{21}$$

2.5.2. Wavelet Edge Detector Implementation

Wavelet edge detection can be demonstrated as in Figure 2. Each process and the corresponding output will be discussed in the following sections.

2.5.3. The Smoothing

We smooth the image and detect the sharp variation points from their first order derivative. The smoothing function is any function whose integral is equal to 1 and that converges to 0 at infinity. One can choose smoothing function equal to a Gaussian.

$$\theta(x,y) = \frac{1}{2\pi\sigma_x\sigma_y} e^{-\frac{1}{2}\left(\left(\frac{x-\mu_x}{\sigma_x}\right)^2 + \left(\frac{y-\mu_y}{\sigma_y}\right)^2\right)} \tag{22}$$

We assume $\sigma_x = \sigma_y$; so the rotation variation is ignored i.e. $\mu_x = \mu_y = 0$ [14].

By convolving the image with smoothing function we removes small signal fluctuation, then sharp variations will only be detected. The value of σ can be changed from image to image according to the amount of edges and detail. Linear filtering of an image is accomplished through an operation called convolution. In convolution, the value of an output pixel is computed as a weighted sum of neighboring pixels. The matrix of weights is called the convolution kernel, also known as the filter. In general case the wavelet domain in 2-dimension has R^4 (Four dimensions) namely (x, y, scale and rotation). At each step, the image is convolved with a wavelet to obtain the coefficients. It is then smoothed with a Gaussian. Both the wavelet and Gaussian filtering are done using separate 1-D filters vertically and horizontally.

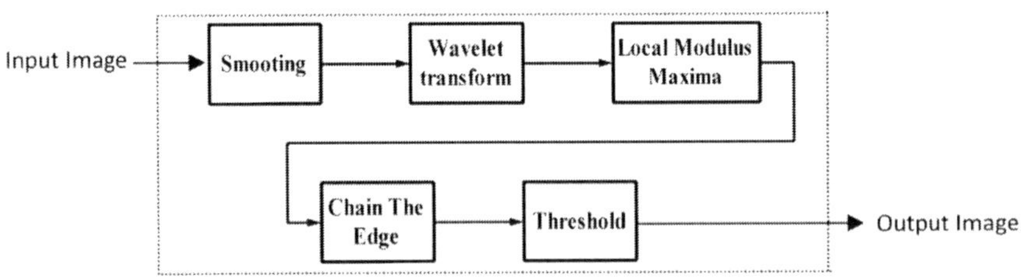

Figure 2. Standard wavelet edge detection method.

2.5.4. Local Modulus Maxima

In order to compute the local maxima of the modulus of the image in DWT we need to find: 1—the modulus and 2—the angle of the DWT at each scale. The modulus maxima are then defined as the points of the modulus images that are larger than the two neighbors whose positions are in the direction indicated by the angle value at the corresponding location. Now let's know how we can get first the modulus of the image:

$$M_s f(x,y) = \sqrt{\left|W_s^1 f(x,y)\right|^2 + \left|W_s^2 f(x,y)\right|^2} \quad (23)$$

Then we get the angle of the DWT. There are only four possible directions when describing the surrounding pixels: 0 degree (in the horizontal direction), 45 degree (along the positive diagonal), 90 degree (in the vertical direction), or 135 degree (along the negative diagonal). Now from the modulus image and the angular DWT we can find the local modulus maxima.

2.5.5. Chain the Edge

The edge points can then be gathered together into chains. For point that is a maximum, we can join it with the point closest to it if they have similar angles. They thus define the multi-scale image edges.

2.5.6. Thresholding

Although the maxima of the modulus of the discrete wavelet transform are a good approximation of the edges in an image, even in the absence of noise often there are many false edges as can be seen. Therefore a criterion such as thresholding must be determined to separate the "true" edges from the "false" edges.

3. A PREPROCESSED EDGE DETECTION APPROACH

Figure 3 shows schematic diagram for the proposed preprocessed edge detection method. In this method the image is processed and the resulting one is applied to standard edge detection method for the objective to enhance the final edge detection method performance.

The proposed preprocessing involves computation of the histogram, finding out the total number of peaks and suppressing irrelevant peaks. From the intensity values corresponding to peaks, threshold values are obtained. From these threshold values, optimal thresholds are calculated using the

Otsu method for multilevel image thresholds [16] . This preprocessing can be implemented in the following steps demonstrated by an image example:

1. Filter the original image Figure 4 using a Gaussian filter as shown in Figure 5.

2. Convert the image to grayscale as shown in Figure 6.

3. Use an averaging mask over the image and resize the image. To limit computation time, we resize the image to have 256 rows and columns that are proportion to it i.e. the original aspect ratio is maintained as shown in Figure 7.

4. A histogram of the image is generated. Thus, the result is of the form [intensity level, number of pixels with that intensity level] as shown in Figure 8.

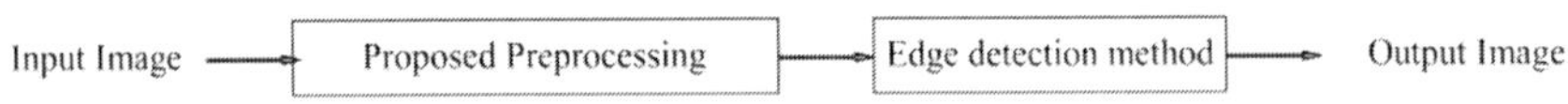

Figure 3. Proposed technique for edge detection.

Figure 4. Original image.

Figure 5. Gaussian filtered image.

Figure 6. Gray image.

Figure 7. Resized image.

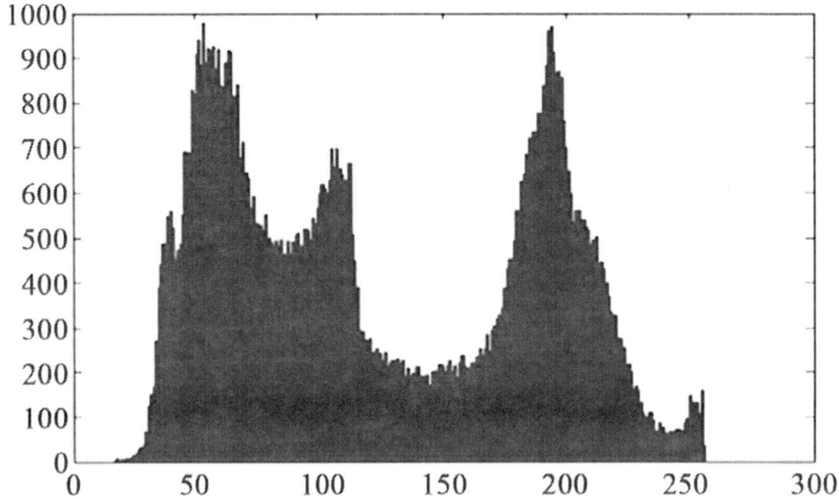

Figure 8. Image histogram.

5. From the histogram data, create a set of values from which the local peaks are counted 58 elements.

6. From the above set, create a set where the peaks that are within 1% of the maximum value (of the number of pixels corresponding to the same intensity level) are counted and others are discarded 57 elements. This factor has been statistically obtained from assessing many images [17] .

7. From the set created in the previous step, remove the peaks that correspond to intensities that are within 20 levels of each other 7 elements. The number 20 is again found effective after experiments by [17] . Now, we have the number of peaks which are thresholds that we can consider from the image. They represent the significant peaks in the histograms shown in Figure 9.

8. Using the threshold values above, find out the optimal threshold using Otsu's method; Thresh = [53 74 98 128 162 190 218].

9. Based in the thresholds obtained from Otsu's method, segment the resized image as shown in Figure 10.

4. SIMULATION RESULTS

This section presents a comparison between different edge detection models. For this purpose MATLAB 8.1 (R2013a) program has been adopted to investigate the difference between edge detection models. Namely, BSD (Berkeley Segmentation Dataset) images [18] and respective ground truths are used for experimentation. It has been found that the best model is wavelet edge detection model. This comparison between Roberts, Prewitt, LoG, Canny and Wavelet edge detection methods is done using ground truth of images. The performance para- meters used are Performance Ratio (PR), Peak Signal to Noise Ratio (PSNR), False Alarm Count (FAC), Miss Count (MSC), and Figure of Merit (FoM). PR is defined as the ratio of true to false edges [19] .

$$PR = \frac{\text{True Edges}\left(\text{Edge pixels identified as Edges}\right)}{\text{False Edges} + \text{Edge pixels identified as non} - \text{edge pixels}} \times 100 \quad (24)$$

The FoM is defined by

$$\text{FoM} = \frac{1}{\max\left(I_A, I_I\right)} \sum_{i=1}^{I_A} \frac{1}{1 + d_i \propto^2} \tag{25}$$

where, false edges are the non-edge pixels identified as edges, I_A is the detected edges, I_I is the ideal edges, d is the distance between actual and ideal edges and $\propto$ is the penalty factor for displaced edges. (PR, PSNR, FoM), higher the value of these parameters better the result is. (MSC, FAC), lower the value of these parameters better the result is. Table 2 summarizes the performance measures of standard edge detection methods men- tioned in this paper with and without preprocessing approach. We refer to the detection method without preprocessing as standard method, and with preprocessing as preprocessed method.

As shown in Table 2, the preprocessed edge detection methods provide better performance measures than the standard ones. However, the preprocessed edge detection methods take more computation time due to the preprocessing. Figure 11 shows the original image used in simulation and its ground truth and the output of standard and preprocessed methods mentioned in Table 2.

53	977
78	550
105	697
126	244
193	970
227	167
254	159

Figure 9. Otsu threshold.

Figure 10. Otsu image.

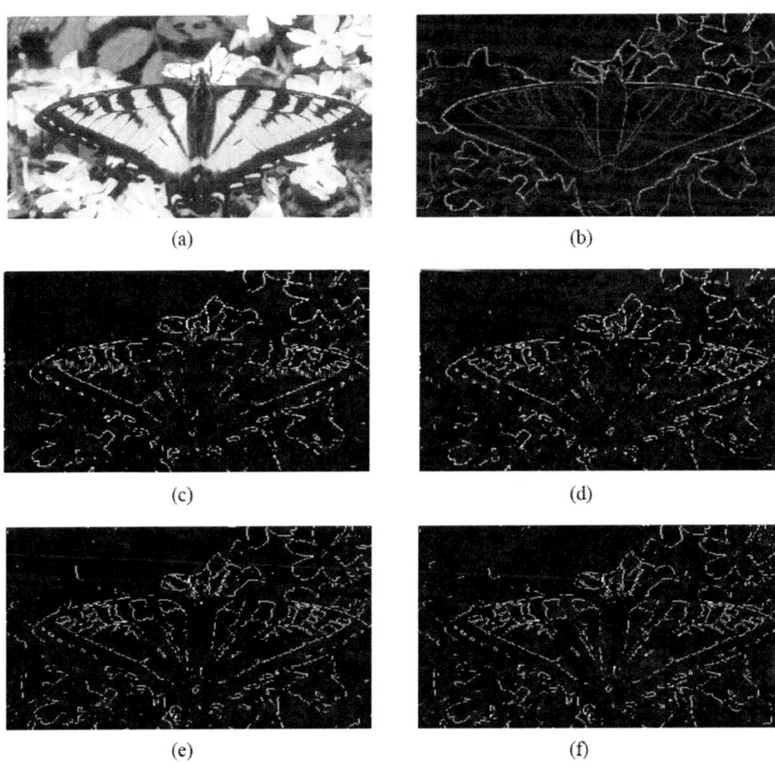

(a) (b) (c) (d) (e) (f)

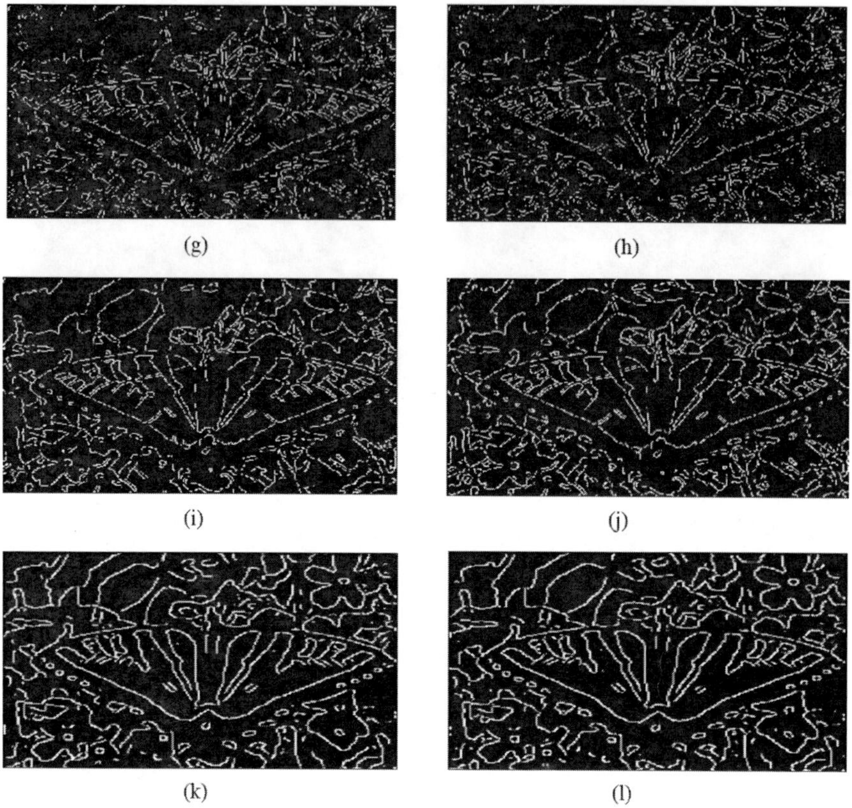

Figure 11. Comparison between various edge detection models (a) Original image; (b) Ground truth; (c) Robert; (d) Proposed Robert; (e) Prewitt; (f) Proposed Prewitt; (g) LoG; (h) Proposed LoG; (i) Canny; (j) Proposed Canny; (k) Wavelet; (l) Proposed wavelet.

Table 2. Comparison between the proposed preprocessed and standard edge detection methods.

Edge Detection Method		PR	PSNR db	FoM	FAC	MSC
Roberts	Standard	9.8844	17.2154	0.083204	1482	3934
	Preprocessed	9.9116	17.2153	0.086603	1478	3926
Prewitt	Standard	9.5487	17.2104	0.094635	2004	3922
	Preprocessed	9.6984	17.2106	0.095899	1954	3921
Zero Cross (LoG)	Standard	11.6908	17.2118	0.12455	4792	3952
	Preprocessed	11.3603	17.2120	0.12633	5086	3960
Canny	Standard	17.1933	17.2235	0.11672	4948	3969
	Preprocessed	17.2174	17.2240	0.11936	6116	3957
Wavelet	Standard	63.5240	17.2272	0.047461	0	2024
	Preprocessed	63.5461	17.2272	0.048007	0	2012

5. CONCLUSION

Mohamed Abo-Zahhad, Reda Ragab Gharieb, Sabah M. Ahmed, Ahmed Abd El-Baset Donkol In this paper, we have presented a preprocessing approach in order to enhance the performance of commonly used edge detection methods. The preprocessing approach consists of computation of the image histogram, finding out the total number of peaks and suppressing irrelevant peaks. From the intensity values corresponding to relevant peaks, threshold values are obtained. From these threshold values, optimal multilevel thresholds are calculated using the Otsu method, then multilevel image segmentation is carried out. Finally, a standard edge detection method can be applied to the resultant segmented image. The preprocessing approach has been summarized in steps and demonstrated by an image example. Simulation results have shown that the preprocessing approach is used with standard edge detection methods to enhance their performance. Results have also shown that wavelet edge detection with the preprocessing approach provides the superior performance among other standard methods with and without preprocessing approach.

REFERENCES

1. Mukesh, K. and Rohini, S. (2013) Algorithm and Technique on Various Edge Detection: A Survey. International Journal (SIPIJ), 4, 65-75.

2. Canny, J. (1986) A Computational Approach to Edge Detection. Transactions on Pattern Analysis and Machine Intellgence, 3, 679-697.

3. Rufeil, E., Gimenez, J. and Flesia, G. (2012) Comparison of Edge Detection Algorithms on the Undecimated Wavelet Transform. CLAM 2012, Latin American Congress in Mathematics, Argentina.

4. Marr, D. and Hildreth, E. (1980) Theory of Edge Detection. Proceedings of the Royal Society of London. Series B, Biological Sciences, 207, 187-217.

5. Urkowitz, M. (1969) A Nonlinear Edge-Detection Technique. Proceedings of IEEE, 814-815.

6. Mallat, S. (1998) Wavelet Tour of Signal Processing. Academic Press.

7. Xu, Y., Weaver, J.B., Healy, D.M. and Lu, J. (1994) Wavelet Transform Domain Filters: A Spatially Selective Noise Filtration Technique. IEEE Transaction on Image Processing, 3, 747-758.

8. Brian, M. and Swami, A. (1999) Analysis of Multiscale Products for Step Detection and Estimation. IEEE Transaction on Information Theory, 45, 1043-1051.

9. Bao, P. and Zhang, L. (2003) Noise Reduction for Magnetic Resonance Images via Adaptive Multiscale Products Thresholding. IEEE Transaction on Medical Imaging, 22, 1089-1099.

10. Rani, V. and Sharma, D. (2012) A Study of Edge-Detection Methods. International Journal of Science, Engineering and Technology Research (IJSETR), 1, 62-65.

11. Gonzalez, R.C. and Woods, R.E. (2002) Digital Image Processing. 2nd Edition, Prentice Hall, Upper Saddle River.

12. Saluja, S., Singh, A.K. and Agrawal, S. (2013) A Study of Edge-Detection Methods. International Journal of Advanced Research in Computer and Communication Engineering (IJARCCE), 2, 994-999.

13. Raman, M. and Aggarwal, H. (2009) Study and Comparison of Various Image Edge Detection Techniques. International Journal of Image Processing (IJIP), 3, 1-12.

14. Bella, B., Tawfik, S. and EL-Gazzar, M.M. (2013) Multiscale Edge Detection Using Wavelet Transform Compared to Other Methodologies. International Journal of Mathematics and Computer Applications Research (IJMCAR), 3, 103- 114.

15. Seema, B.K. (2013) An Edge Detection Approach Based on Wavelets. International Journal of Engineering Research & Technology (IJERT), 2, 2836-2840.

16. Otsu, N. (1979) A Threshold Selection Method from Gray Level Histograms. IEEE Transactions on Systems Man Cybernet, SMC-9, 62-66.

17. Nain, N., Jindal, G. and Jain, A. (2008) Dynamic Thresholding Based Edge Detection. Proceedings of the World Congress on Engineering, London, 2-4 July 2008.

18. Martin, D., Fowlkes, C. and Tal, D. (2001) A Database of Human Segmented Natural Images and Its Application to Evaluating Segmentation Algorithms and Measuring Ecological Statistics. ICCV Vancouver, 1-8.

19. Khaire, P.A. and Thakur, N.V. (2012) A Fuzzy Set Approach for Edge Detection. International Journal of Image Processing (IJIP), 6, 403-412.

Index